BECOMING A GARDENER

BECOMING A GARDENER

What Reading and Digging Taught Me About Living

CATIE MARRON

Illustrations by
All the Way to Paris

Photographs by
William Abranowicz

HARPER DESIGN
An Imprint of HarperCollins Publishers

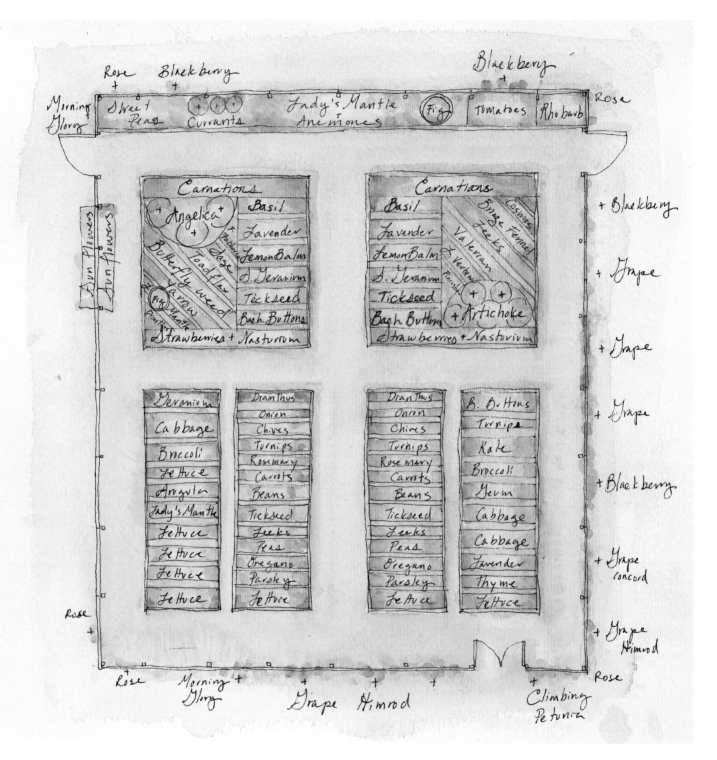

CONTENTS

INTRODUCTION

If you have a garden and a library
you have everything you need.

CICERO

Gardens have mattered deeply in people's lives ever since Eve ate the apple from the tree. Cicero's words, written in his native Latin more than two thousand years ago, still resonate profoundly today. For centuries, gardens and books have fulfilled our human need to enrich both our minds and our souls. It's elemental.

In early 2017, our family came upon a house in Connecticut by chance, less than an hour's drive from our lifelong New York City home. We fell in love with the rolling land that unfurled around it in every direction and the secluded lake at the property's edge. Don, my husband of thirty years, and I had often dreamed of having a home big enough for us and our two children, William and Serena, and future generations of our family. This house fit the bill. It was well designed and, with some needed renovations, paint, and our furniture, it came to look like us. Yet I couldn't shake the sense that I didn't belong there, that I was living in someone else's house. I tried everything to feel comfortable. I even burned sage and applied the principles of feng shui—to no avail.

About a year later, still unsettled, a thought occurred to me: To feel rooted, I had to put down roots. Literally. I found myself channeling Cicero and thinking of gardens. Over the years, I've built a small library of gardening books and nurtured the idea of creating a garden: a space of my own where

I could work in the dirt, be involved with nature, and produce the flowering plants of my imagination. To turn our house into a home, I'd root myself to the land, which is what drew me to the spot in the first place. Perhaps that's because of what author Penelope Lively, whose book *Life in the Garden* is in my collection, thinks is our primeval need: the urge to be outside. It certainly seemed as good a time as any to attempt to become a gardener.

Obviously, to become a gardener, I needed to garden. I had a rough idea of what I wanted—something that fit naturally with my image of the New England countryside—but I didn't know how or where to start. I decided to give myself eighteen months, thinking this would allow me enough time to design the garden and watch a full year's plant cycle. Little did I know how much the world would change in that time, both globally and for me personally. What began as a desire to feel rooted in a new home with my family became something else entirely.

In the beginning, I thought all I needed was gardening advice—basic information on how to lay out a garden and what plants to grow. As it happened, I needed and discovered so much more. I turned to my library of gardening books for help. Looking at them, I realized that my favorites were written by some of the world's most brilliant fiction and nonfiction writers who also happened to be gardeners. As a metaphor, gardening is, in Eudora Welty's words, "akin to writing stories." Both books and gardens give our imaginations a chance to roam and create our own private worlds. They let us escape time, entertain us, and offer pleasure and beauty. Certainly a garden is a glorious place to read. It is rich with life lessons—ample ground for any writer to explore.

I read (sometimes reread) book after book, hoping to uncover some well-hidden secret on what it means to be a gardener, and how to become one. In short order, I read that gardens give enduring happiness, offer hope, teach patience and tranquility, and provide sanctuary and beauty. Beverley Nichols writes that a garden is "a place for shaping a little world of your own according to your heart's desire." According to Anna Pavord, "gardening slows you down, masks worries, puts them in proportion. A garden teaches you to be observant and how to look at things. You become less inclined to leap to quick conclusions. Or to jump on the latest bandwagon. A garden hones your senses." Monty Don thinks that "gardens heal." He writes, "When you are sad, a garden comforts. When you are humiliated or defeated, a garden consoles. When you are lonely, it offers companionship that is true and lasting. When you are weary, your garden will soothe and refresh you."

I soon discovered that there are as many different personalities of writers as there are gardeners but garden writers, in particular, share a common characteristic: They're opinionated. Yet, just as easily, they trade helpful hints with one another as argue. Imagine if what these writers said about gardens was true, how much I'd benefit. Over the course of the next several months, I wrote down memorable quotes and words of advice until I had enough wisdom to compile an anthology. I had endless questions: Where do gardeners get their inspiration? What is the meaning of gardens in their lives? How will I know when I'm a gardener? What exactly is the difference between straw and hay?

And then I read Jamaica Kincaid's *My Garden (Book)*: She writes, "I started to plant things—this is not the same thing as being a gardener— when, to celebrate my second Mother's Day, my husband, on my daughter's behalf, gave me some packets of seeds (I only remember delphiniums and marigolds), along with a rake, a hoe, and a digging fork, all bought from the Ames department store in Bennington, Vermont. I went outside, dug up the yard, and put the seeds in the ground. The skin of my right forefinger split, the muscles in the back of my calves and thighs were sore, and the digging fork broke; for all that, the seeds did not germinate. . . . Since then I have become a gardener."

I felt exactly like that. The best I knew I could ever realistically achieve in eighteen months was to become a novice . . . and that, I decided, would be fine. Over the next year and a half, I made lots of beginner's mistakes, but also experienced enough success to keep me going.

This book is about my education from dreaming to doing. Before I ever planted my first seed, these writers, whom I now think of as my literary mentors, gave me direction and prodded me forward when I felt overwhelmed and lost. Even though many of these authors were or are experts, they all had to start somewhere.

I would eventually discover another circle of mentors—hands-on, real-life gardeners—who taught me the right way to sow and why it's important to understand soil. By digging deeper into my reading and working in the garden, both with my mentors and by myself, I learned a new way of being rooted, as Michael Pollan writes, "in the endlessly engrossing ways that cultivating a garden attaches a body to the earth."

When I began this project, a friend told me I was trying to cram a lifetime of study into months. She was right: A gardener's education is never done. But I've realized I'm not taking a test to prove anything and, besides, there is no test. For as many different answers as there are to what a garden is for, there are also multitudes of reasons why we garden.

I've always loved the saying, "Shoot for the moon. Even if you miss, you'll land among the stars." Why not try? I had a garden, I had books. This is my wholehearted attempt.

I. GETTING MY BEARINGS

Nature is such a physical pleasure that it is tempting to sink into it entirely. . . . I'm sure it made me a gardener, this self-soothing alertness, this urgent need to look. Maybe it makes writers, too.

CHARLOTTE MENDELSON

IMAGINATION AND MEMORY

How appealing it would be to sink into nature entirely. Or, even better, to have nature sink into me. When we first arrived at our property in Connecticut, I was awestruck by the pure, timeless sense of nature. Trees and grass, water and woods. Sitting on the old New England stone wall next to the house, I could picture the land in colonial times. Its simplicity made that easy.

Creating a garden requires imagination. Many writers encourage this, with American historian Alice Morse Earle believing that "half the interest of a garden is the constant exercise of the imagination." Once I started imagining a garden of my own, all sorts of memories came swiftly back.

As a child, I lived in my imagination and in trees: My earliest memory is of looking out of my bedroom window at a pale pink crabapple tree, the moon shining behind it in a midnight blue sky. I thought I'd be quite alone in admitting that I climbed and swung in trees up until college but then I read that Eleanor Perényi climbed them "until well past the age when it was decent to do so." Her children put an end to it. She captured my feeling completely when she wrote "to climb up, up, among the leaves, beyond the reach of intervention, must be the oldest, most joyous instinct in the world, the next best thing to flying." Like her, in the trees, I was in another world.

Thoughts about flowers, plants, and gardens often come straight from memories of our childhood experiences. As Robin Lane Fox writes: "Flowers and gardens from the past stay with us to the end." The enchantment of flowers, gardens, and nature are very real, but their magic can be discovered in the pages of a book, too. The magical kingdoms in my childhood books were the first times in my life I encountered special gardens. In several favorites that I first read as a child, and then as an adult to my children, the garden was a central character.

A world of my own is where I went, again and again, whenever I ducked under the covers and read Frances Hodgson Burnett's *The Secret Garden*. Thoroughly enchanting, it is a story about the importance of nature to the human spirit. Mary, a young orphan—along with Colin, a cousin she didn't know she had—discovers an abandoned secret garden on her uncle's Victorian country estate. The garden's power is so transfixing, it arouses Colin from his bed to see spring arrive: "And the secret garden bloomed and bloomed and every morning revealed new miracles." Mary sets out to restore the garden and, in the process, discovers it has magical powers: Flowers grow before Mary's and Colin's eyes; dogs and robins seem to talk. With its ancient sprawling roses, apple trees trained flat against the wall, and ivy covering up a hidden door, the garden is a private space of mystery and healing. Burnett writes, "Sometimes since I've been in the garden I've looked up through the trees at the sky and I have had a strange feeling of being happy as if something were pushing and drawing in my chest and making me breathe fast. Magic is always pushing and drawing and making things out of nothing. Everything is made out of Magic, leaves and trees, flowers and birds, badgers and foxes and squirrels and people. So it must be all around us." How lovely that would be.

The Tale of Peter Rabbit is another wonderful book in which the garden plays a major role. Beatrix Potter first conjured up this mischievous rabbit to entertain and cheer up her former governess's son who was often ill. While Potter's illustrations are as beloved as her tales, she gives us only tantalizing glimpses of Mr. McGregor's vegetable garden: a cucumber frame on the corner of the page,

ABOVE: *An illustration from Beatrix Potter's* The Tale of Peter Rabbit, *1902.*
OPPOSITE: *Ludwig Bemelmans's illustration of Luxembourg Gardens, from his book* Madeline, *1939.*

a few old pots of geraniums, a cabbage here or there, and a gate that looks like it creaks. And, yet, you do see the garden. You can picture it through the Westmorland mist as if you're standing next to McGregor himself.

As adults, we can return to a childhood favorite not only to revive it in our minds, but also to discover something new. I've recently reread some of these stories, hoping to remember and perhaps be inspired by what I loved most about these enchanted worlds. From the opening line of every book in the series—"In an old house in Paris that was covered with vines lived twelve little girls in two straight lines"—to the last—"That's all there is, there isn't anymore"—Ludwig Bemelmans's lively stories and spirited illustrations have made the Madeline books legendary classics. Growing up, I longed to develop Madeline's spunk and fearless nature. In fact, I had a print of one of the illustrations from *Madeline*, the first book in the series, on my bedroom wall for my entire childhood: our heroine "pooh-poohing" the tiger at the zoo, while Miss Clavel and the other girls look on with fear and astonishment.

The book doesn't take place in a garden, but it celebrates a special one. Over the years this part of the story had escaped me, but I loved rediscovering that Madeline visits Luxembourg Gardens, as it's one of my favorite places of all time. Gardening and reading, memory and imagination—all weave together in books and in life. Perhaps the power of literature and nature are one and the same, calling on gardeners and readers to recruit both their imaginations and their memories to cultivate special places of their own.

INSPIRATION FAR AND WIDE

As I tapped into my memories, I thought of real gardens I've loved. Whenever I travel, I seek out gardens. I wondered if my favorites might give me ideas for mine. As it turned out, they are simply too different from any garden I could make. But their beauty certainly inspires me, and, like my books, they've spurred me on.

The word "paradise" is derived from ancient Persian, meaning "an enclosed garden." Over time, it has come to describe any place of deep beauty and pleasure. I've used this word only once and that was to describe Italy's Garden of Ninfa, perhaps my favorite place in the world.

Before falling in love with Ninfa, my favorite place had been the Luxembourg Gardens, the large park on Paris's Left Bank that Ludwig Bemelmans brought into Madeline's world. While I now feel as if I know it almost as well as the back of my hand, it took me years to fully grasp its scale. I visited it on my first trip to Paris when I was twenty-three. Walking past the Luxembourg Palace, which today houses the French Senate, on a brisk and sunny December morning, I suddenly came upon an expansive gravel courtyard, with a large basin where children were sailing model sailboats. A surprising number of people were out, most of them gathered around the boat pond, soaking up the warmth of the sun. As I watched the scene before me, tears came to my eyes. Something about the contrast between the formal, beautiful setting and its natural, everyday humanity was deeply poignant. Each time I've visited Paris since then, I've made a pilgrimage to the Gardens, even if it meant stopping en route to the airport. I've grown up with this park; I used to visit it alone, then with my husband, and together we introduced our children to the grounds, which have English and French gardens and an old-fashioned merry-go-round where children reach out to capture a golden ring. It was always my dream spot—until I went to Ninfa.

Once an ancient medieval town tucked into the base of a small mountain range in central Italy, Ninfa had, over the course of several centuries, been host to many wars that reduced the town to ruin. It began coming back to life in the early twentieth century when the first of three generations of the Caetani family devoted themselves to transforming it into the beacon of otherworldly beauty it's known for today. Those who are fortunate enough to have visited Ninfa often describe it as the most romantic garden in the world. Hugh Johnson wrote, "It is over fifteen years now since I first visited Ninfa but the vision of this dreaming garden-ruin . . . has never left me. It stuck in my mind as a place so beautiful and extraordinary that even the potent word garden could hardly stretch to encompass it."

My family and I went there in early March about twelve years ago on a day when you could feel spring in the air. A friend had insisted we'd love it and organized our visit while we were on vacation. As we drove farther away from Rome, I worried that I'd dragged my family away from a day they'd prefer to spend exploring the city. But when we arrived, we were greeted by a large, friendly dog, who became an instant friend and welcomed us into the garden. I immediately understood why ancient Italians named Ninfa after a nymphaeum, a temple dedicated to nymphs. It looked and felt as if we'd stepped into the pages of a fairy tale. We'd have gotten lost if it weren't for Ninfa's deeply gracious director and our guide, Lauro Marchetti, who is devoted to Ninfa's conservation and protection.

The garden is unlike any I've ever seen—dreamy doesn't even begin to describe it. Flowering trees seem to run wild everywhere. Roses are abundant, many climbing up the medieval ruins. Colors are vibrant: purple clematis, white

PAGES 34–37 AND 39: *The Garden of Ninfa.*

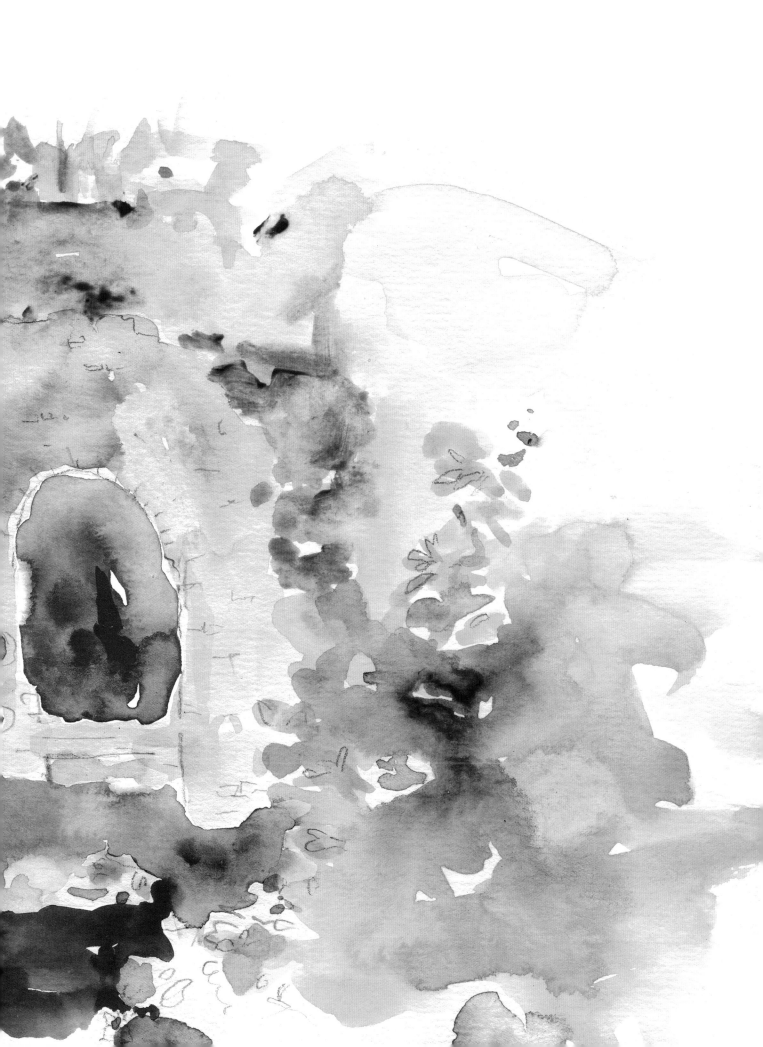

magnolia, golden yellow trumpet vines, and pale pink cherries. Most of all I remember happy springtime shades of green. A stream, which flows from a spring deep in the mountains, had the clearest water I've ever seen. Brilliant green moss stretched along its banks and flowed beneath the water like the strands of a mermaid's hair.

In a small hut by the stream, the Italian writer Giorgio Bassani wrote *The Garden of the Finzi-Continis*, an exquisite and moving book. The movie adaptation, directed by Vittorio De Sica, is one of my top five favorite films. It is both other-worldly and haunting, and ultimately deeply sad. As I have a lot of Italian blood, it was powerful to make the connection between the garden in the film and Ninfa, its real-life inspiration.

When we came to our final destination, a stone villa on the property where Franz Liszt practiced the piano, there was an orange tree already in bloom with luscious fruit. Lauro invited us to pluck one and also gave us fresh juice. Deeply refreshed, we drove back to our hotel in bliss. Don and I spoke about Ninfa many times thereafter, and what the experience meant to our family: If only for a few fleeting hours, we'd tasted paradise.

When I first visited Spain I was in my early thirties and joined my husband on a business trip. Each afternoon while he worked, I walked to the Madrid Royal Botanical Garden, which was close to our hotel and next door to the Prado Museum. Spread across three tiered terraces, the garden was created in the late eighteenth century to hold King Carlos III's collection of exotic plants in several of its greenhouses. It feels larger than its nineteen acres, perhaps because of its winding paths and mature trees. Its herbarium is the largest in Spain. I didn't venture inside the greenhouses, however, preferring to wander outside. I recall twice sitting on a bench surrounded by tall, dark evergreens, and staring at the inky blue sky, the kind that portends a thunderstorm, which mesmerized me. The combination was dramatic, creating a rather brooding feeling, making it all the more captivating. Not long thereafter, I read Roy Strong's *Garden Party*, and felt flattered that he too loved the Madrid garden. Rereading it now, I enjoy seeing that he also included Ninfa as one of his favorite gardens.

Le Jardin Plume.

The Isabella Stewart Gardner Museum in Boston is another garden I visited often when I was in college outside of the city, despite it being very much off my beaten path. On the final leg of my one-hour trip, I'd exit the subway and walk across a large field to arrive at the museum. Built in the early twentieth century to resemble a fifteenth-century Venetian palace, it had an inner courtyard garden like those of Renaissance villas. The courtyard was always filled with fresh flowers. Looking back, it's no wonder I found it alluring: It gave me the chance to be alone and absorb the lush Italianate beauty of the garden. The museum's collection was wonderful, but it was the garden that again and again drew me back.

If you could travel to gardens you've only dreamed about, where would you go? If I were planning this adventure, I'd opt for a weekend holiday each season of the year, and these are the gardens I'd visit.

First stop: Le Jardin Plume in Normandy, France. Its creation, still ongoing like all great gardens, began in 1996 when Patrick and Sylvie Quibel first arrived in the area and discovered a flat stretch of land previously occupied by roaming sheep. Today, the garden is a fascinating mix of classic French formal garden traditions, such as the allées where you can see the influence of Versailles, with a more pared-down, modern layout and loose plantings. Plume means "feather" in French, and the Quibels have chosen plants for their feather-light qualities, wanting them to be in harmony with the flight of butterflies. The garden seems both ephemeral and rooted. The Quibels used grasses before they became popular, and they and other feathery perennials often blow in the breeze while being protected by tall box hedges pruned precisely into the shape of undulating waves. It's at peak beauty in the fall, so I'd head there then.

In the winter, I'd go to the Netherlands to see one of Piet Oudolf's gardens in his native country or, if that wasn't possible, certainly another one of his gardens somewhere else in the world. Oudolf is considered by many to be one of the most innovative landscape designers of the past twenty-five years.

His gardens are naturalistic in form and design, like an idealized version of nature. Oudolf creates an ethereal wildness, yet one where he's in complete control, taking action—something he calls "editing"—whenever necessary. For the past thirty years, he and his wife, Anya, have lived at Hummelo, patiently developing it as a testing ground for new plants and ideas. Oudolf chooses plants more for their shape and texture than for their blooms—creating a garden of beauty through all four seasons rather than a showstopper for one or two. Like the Quibels, Oudolf was a pioneer in the use of grasses, which he planted in billowing masses to create a blurred, hazy background against which other plants can stand out.

I know one of his most famous gardens very well—the High Line in New York City—and have studied its plans. Its flow of plants, so many more in even a square foot than might meet the eye, looks spontaneous yet the plan is meticulously thought out. My favorite time to visit the High Line is on a late winter afternoon when the perennials have faded but the red berries pop through the grasses, which feel so airy and free and shine a golden brown in the slanting, crisp afternoon light. It's rare to be in the middle of New York City and also have a sense of escape and solitude.

On a hot summer's day, I'd go to Dungeness, Derek Jarman's garden in Kent on the English Channel. As lush as Le Jardin Plume and Hummelo are, Dungeness is rough. Modest in size, the garden rises from the stark, flat, coastal landscape that juts into the Channel and surrounds Jarman's house, a small fisherman's cottage. There are no tickets to buy, no restrictions, no reservations. Visitors can wander around Jarman's yard whenever they like.

Jarman created the garden in the final years of his life, and documented its evolution with his friend Howard Sooley, who describes the garden as stretching to the horizon in all directions. A distant nuclear power station hums by day and glows by night. In the summer, brightly colored wildflowers and all sorts of poppies, scattered in the seaside brush, and stones wrap around the cottage—a striking sight as the cottage is painted black with bright yellow trim. It's a study in extremes.

My last stop would be closer to home: I'd head to Thomas Jefferson's Monticello garden during spring planting season. My family and I actually visited his house late one afternoon at the beginning of a new year, and I vividly recall the peacefulness of Virginia's Blue Ridge Mountains at dusk. In the hush of winter, with no one else around, I walked into Jefferson's bedroom just when the candles would have been lit and imagined him working there. At the time, I didn't know much about Monticello and never thought to seek out the garden. Little did I know then how much I would come to refer to its rows of lettuces and what Jefferson called patches of peas as I planned out my own garden. I would like to visit when the vegetables start to ripen and mature.

Gardeners get multiple doses of pleasure: They get to enjoy the garden in its present form while also imagining what it will become three months hence while simultaneously reminiscing about its past. In this way, memory and imagination are inextricably linked—to themselves and to the garden.

SINKING INTO NATURE

As I became better acquainted with our real-life spot in Connecticut, quickly developing a routine loop for my morning walk with our dog, I noticed more and more and wanted to know so much more. The first step in doing that was to focus outward as well as inward: to look carefully at the natural world around me, both at our house and within the density of New York City—everywhere.

Once I started looking, it became, as Charlotte Mendelson writes, an "urgent need." Through looking closely, I learned to observe. The more I observed, the more I understood what I was reading and seeing, and the more I read, the more I looked. A continuous happy cycle.

Philosopher and author Alain de Botton taught me an invaluable lesson that applies to flowers as much as to postcards. His idea is as follows: "Our culture sees them as tiny, pale shadows of the far superior originals hanging on the walls

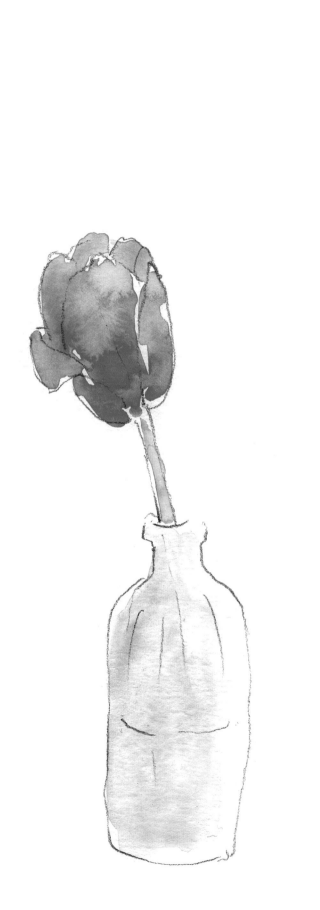
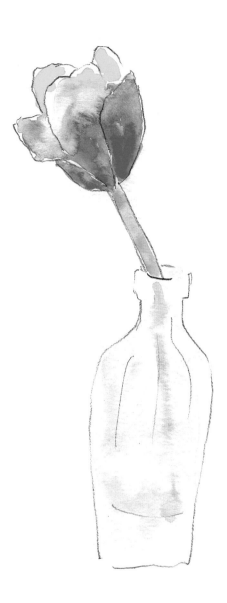

a few metres away, but the encounter we have with the postcard may be deeper, more perceptive and more valuable to us, because the card allows us to bring our own reactions to it. It feels safe and acceptable to pin it on a wall, throw it away or scribble on it, and by being able to behave so casually around it, our responses come alive."

In keeping with this idea, I often put one flower in a small bottle on my desk and watch its cycle. The changes that happen over several days are astounding. I've seen the petals of a tulip almost double in size, a peony change from hot pink to pale coral, and a rose cling to its fading bloom. I've not only learned how to look more keenly, but equally important—and in much the same way that a postcard of a great painting turns it into something relatable, I've learned that the garden is hardy, it can withstand changes and mistakes. The richness of this new dimension in my life was unexpected—an early bonus. After a while, looking closely became an ingrained practice. I'd frequently stop to examine a surprising color combination or a plant that looked like something I recognized but wasn't. I felt like Beth Chatto when she visited Cedric Morris's famous garden and wrote, "But that first afternoon, there were far too many unknown plants for me to see, let alone recognise. You may look, but you will not see, without knowledge to direct your mind. As you become familiar with more plants and plant families your eye will pick out the unfamiliar ones and so add to your pleasure and knowledge." You read to understand meaning. You look to see what you like. And I hoped my garden would bring both to fruition.

Moving and growing as it does with the seasons, a garden can never be pinned down. Like the flower I observe on my desk, the garden is in a perpetual state of change—its various plants are always in different stages of the life cycle. In this sense, garden time is amorphous; it moves with the ease of flowers blowing in the wind. A garden represents the memory of yesterday, the immediacy of today, and the imagination of tomorrow—life in full.

II. GROWING PRINCIPLES

Tell me about the garden you're dreaming of,
and I will tell you who you are.

JEAN-JACQUES ROUSSEAU

GENIUS OF THE PLACE

For many years, my dream garden was the one Don and I had in Long Island where we bought our first house together. I remember visiting it on several early Saturday mornings before we moved in, a few years before we had children. The property, not having been lived in for several years, looked mysterious in the morning fog. The rhododendrons along the driveway were so thick that my car almost had to push through them. I walked around as if in a real-life version of *The Secret Garden*. The land was beautifully overgrown with wild and ancient-looking trees, some of their trunks covered with ivy, and strangely shaped evergreens. The previous owner had planted many unusual bushes and trees, among them two dense camellias that still bloom each year. The topiary, a wreck of a privet elephant and a yew duck and rabbit, are still there, perhaps recognized only by us some thirty years later.

When we moved in—before hurricanes and old age struck several of the trees—two large, full maples stood next to the terrace. They'd clearly been there for decades, and we enjoyed their wonderful shade for years, especially when having meals outside on hot, sunny days. They both had a strong branch ideally placed for an old-fashioned swing just like the kind I'd had growing up.

One of my favorite memories of the wonderful times spent there is of my husband and me pushing our children on the swings. When we'd get the timing right, our son and daughter would reach their feet toward one another to touch toes. Fortunately, we caught this on camera, so we'll always be able to look at this magical moment.

The house and the garden have a beguiling spirit all their own. Even when the trees have lost their leaves, the property is still inviting. In the wintertime, it's a snowy wonderland, especially when the snow rests on the limbs of the trees. Our gardener kept our trees healthy for over twenty-five years, and I hoped I helped him by losing myself many summer weekends in the bushes, happily weeding and pruning, two aspects of gardening I knew how to do.

Some years ago, I sought out Miranda Brooks, a very talented landscape designer with a romantic sensibility I love, to restore the feeling of the garden to its original character. Miranda transformed it, turning a lawn and old trees into a breathtaking, lush garden full of peonies, roses, lilacs, wisteria, herbs, and grasses. She even created a hidden cutting garden. I now feel as if I live in a garden, especially in June, when everything blossoms together. I often wished I could shadow Miranda for a few years to see how it's done. As a friend once said to me, "Catie, you used to live in an arboretum. Now you live in a garden."

In Connecticut, before deciding to make a garden, I wanted something different: the simplicity of trees, grass, and woods, easy to maintain and in keeping with my sense of New England. I believe in the concept of a garden echoing the spirit of its surrounding landscape, something Alexander Pope called in the eighteenth century "the genius of the place." Even though it's just across the Long Island Sound, Connecticut is a different world—from the area's agricultural history to the look and feel of its countryside, even the sunlight. In Long Island, the sunlight coats everything in a golden, amber glow. The quality of light is so rare that it's lured artists such as Fairfield Porter, Willem de Kooning, Jackson Pollock, and Lee Krasner to the eastern end of the island for inspiration. In Connecticut, the light seems more white than golden.

I like the simplicity of the rolling hills of Connecticut and the way the countryside evokes images of old New England winters, cold and bleak and gray. The landscape and range of weather, much of which can be quite harsh, reminds me of scenes from early American literature. When I pictured my garden here, I didn't want it to stick out; I wanted it to blend in and be part of the American tradition of gardening and farming, which has deep roots in New England. Our land in Connecticut gave me the perfect opportunity to create a working kitchen garden, something in keeping both with the genius of the place and my deep respect for the surrounding area and its agricultural history. One of the best surviving examples of the American kitchen garden is Jefferson's Monticello so I focused my attention there.

THOMAS JEFFERSON AND THE AMERICAN GARDEN

One of America's most dedicated agricultural innovators, Thomas Jefferson spent many hours working at Monticello, his plantation in Virginia. Besides creating a one-thousand-foot-long vegetable garden, he made other significant contributions to American gardening, the most important being his highly accurate, detailed scientific records and his cross-fertilization of plant species, which yielded new types of cucumbers, zucchini, and melons. He grew 330 varieties of ninety-nine species of vegetables and herbs—some of which were new variations brought to America by his international friends. Jefferson believed that "the greatest service which can be rendered any country is to add an useful plant to it's [sic] culture."

Jefferson was primarily interested in growing food, writing, "I have lived temperately, eating little animal food, and that . . . as a condiment for the vegetables which constitute my principal diet." He also used plants to bring people together. His annual spring pea competition was a popular social function: Whoever harvested the first spring pea held a community dinner that included a feast of the peas. Jefferson grew nineteen varieties of peas and twenty-four varieties of kidney beans! He documented fifty-seven years of work in his Garden Book, a diary of great significance in American history. Still gardening at sixty-eight, he wrote, "But tho' an old man, I am but a young gardener." Further, "I am constantly

in my garden or farms, as exclusively employed out of doors as I was within doors when at Washington, and I find myself infinitely happier in my new mode of life."

Jefferson was a classic American dirt gardener, a distinction that separated him from the Englishman who, according to a nineteenth-century Virginia horticulturist, "prepares his borders while the American digs his holes." However, he wouldn't have realized his vision without the people he enslaved. Slaves dug the now-famous vegetable beds, hauled soil, built fences, and labored daily to maintain the gardens and the fields of tobacco and wheat. Much of Monticello's produce was grown by enslaved families in their own small gardens, probably because Jefferson's efforts were so focused on his experimental ambitions. Slaves could only tend to their plot after a full day's work elsewhere, but the produce they sold allowed them to carve out a tiny piece of independence. They also held on to their heritage by nurturing plants native to Africa such as watermelon and okra.

Today, Monticello, including its famous garden, has been restored to its original plans. Gardeners visit to study its pioneering techniques and to take in the sweeping views of the Blue Ridge Mountains. Peter Hatch, the longtime and now former director of grounds and gardens at Monticello, says that it's "the true American garden: practical, expansive, wrought from a world of edible immigrants."

Alice Waters writes that America is built on the principles of our farmers, and that "our relationship to the land our food comes from is one of the most fundamental relationships of all." The farming tradition is part and parcel of the American identity: The early colonists created gardens that were not only sources of food but also beautiful. Enslaved people held on to their dignity and history by growing food they could eat and sell to others. It was a hardworking approach, which, while initially borne out of survivalist conditions, is one that's continued today. Every garden in America—from Michelle Obama's White House garden to the great farmlands that stretch from sea to shining sea to the victory gardens of World War I and today's homegrown backyard and windowsill gardens—finds its roots in this horticultural pursuit of happiness. Americans are dirt gardeners, a term I use with deep reverence because it is exactly the kind of gardener I want to be.

OPPOSITE: Nathaniel K. Gibbs, *In the Vegetable Garden, Monticello*, 2000.

World War II posters advocating victory gardens. CLOCKWISE, FROM TOP LEFT:
Hubert Morley, Your Victory Garden Counts More Than Ever!, *1945; Unknown artist,*
Paul J. Howard's California Flowerland, *c. 1940; Mary Le Bon,* Dig for Plenty, *1944;*
Unknown arist, Dig for Victory *(New Zealand), c. 1943.*

KEEP CALM AND GARDEN ON

War gardens made their first appearance in America during World War I (1914–1918), when President Woodrow Wilson encouraged Americans to plant vegetable gardens to help prevent food shortages. Postwar, the campaign continued. In 1919, the National War Garden Commission distributed a pamphlet that described gardening as a civic duty with text that read, "Prevention of widespread starvation is the peacetime obligation of the United States. . . . The War Garden of 1918 must become the Victory Garden of 1919." The victory garden—a term coined by Charles Lathrop Pack, head of the National War Garden Commission—quickly became a popular movement in the United States, the United Kingdom, Canada, Australia, and Germany, and lasted through the end of World War II. Americans grew food wherever they could—backyards, abandoned lots, fire escapes—and by the 1940s, more than twenty million victory gardens were planted throughout the country.

Unlike the years of World War I, overall food-production levels throughout World War II were relatively stable. People created victory gardens because they represented a unified patriotic display of solidarity, and corporations were quick to back products made with the victory gardener in mind. By most accounts, victory gardens were largely a transitional stage in American life, with many families favoring postwar processed foods over cauliflower and kale.

Today, particularly in the face of a global pandemic, many Americans have been going back to their roots and planting gardens again. Nurseries have recorded an uptick in sales for staples like squash, onions, cabbage, and herbs. Eager to build their own community-based food security and perhaps as a way to also cultivate something sustainable, people are enthusiastically embracing their cultural heritage as dirt gardeners, learning patience and perseverance in the process, as I have been.

MOON GARDENING

George Washington and John Adams were also early agricultural innovators who advanced various farming techniques. Long before Adams experimented with the compostion of fertilizer, mixing it with mud, lime, and seaweed, Native Americans were developing their own agricultural practices, with many following the principles of moon gardening, which many gardeners still swear by today. It makes sense: As the ocean tides are affected by the gravitational pull of the sun and moon, so too is the water beneath the ground, waxing and waning to a natural rhythm and enriching the plant life accordingly.

Over the course of a twenty-nine-and-a-half-day lunar cycle, the moon goes through four basic phases: new, full, and two quarter phases. The basic principle of moon gardening is to plant aboveground crops during the first two weeks of the lunar month while the moon is waxing, pulling water to the surface of the garden and increasing its moisture. The peak time to harvest is during the full moon when the plants are at their strongest. As the moon wanes over the following two weeks, its gravitational pull weakens: This is the optimal time to plant root crops like beets and carrots as well as to fertilize, transplant, and prune. In essence, this is the best time to prepare the garden for the next phase of the new moon. During the final quarter it's best to avoid planting at all, and work, instead, on improving the soil condition. To adhere to moon-gardening principles, you need the discipline and the ability to literally be in the right place at the right time. This method of gardening acknowledges people's desires to live more in rhythm with nature, a most appealing idea.

I became particularly enamored by the moon during a late summer trip with my son to Scotland. The weather there was unlike any I'd experienced, with changes by the hour. First sun and then rain, and then a dramatic temperature change would bring hail or intense wind. There were rainbows each afternoon. In the middle of one night, I woke up and happened to look outside. The moon was full and the brightest yellow I'd ever seen, beaming through white

clouds in the darkest blue sky I'd also ever seen. It was so magical that I called it the Harry Potter Moon and have thought of it fondly many times since.

One of the pleasures of my garden is to check on it at night when I walk our dog. The garden glows by the light of the moon—in England, nighttime gardens are sometimes called ghost gardens due to their haunting quality. I've resisted the temptation to install electric lighting nearby, thinking I would carry an old-fashioned-looking lantern instead. So far I haven't found the perfect lantern, but the moonlight has worked well as my guide. White flowers look especially beautiful in moonlight—the principal reason Vita Sackville-West planted her legendary white garden at Sissinghurst, their color intensifying against the surrounding darkness, rooted versions of the constellations above. She created this famous garden so that she and her husband, Harold Nicolson, could find their way in the dark from one house to another on their estate. I typically don't like white mixed in with other flower colors simply as a matter of personal preference. To try to understand the allure and value of a ghost white garden, I planted a white moon flower, which comes out only at night. When the full moon comes up over the hill near our property and takes over the black night sky, I can imagine the landscape as it was hundreds of years ago.

ON TYPES OF GARDENERS

I have come to think that there are five types of gardeners: scene setters, plantspeople, colorists, collectors, and dirt gardeners. The distinguished English landscape designer Penelope Hobhouse is the very definition of a scene setter, writing, "As time has passed, I have become less interested in the close-up, in the individual plant . . . and I am more and more interested in the broader picture, the wider landscape." Piet Oudolf is certainly one of the world's most highly regarded scene setters.

Plantspeople zero in on specific plants, and care about their health, looks, and reason for being. French Impressionist Claude Monet, known for rendering the beauty of the garden—his and others—in landscape paintings, was not only a scene setter but also very much a plantsman. He attended the summer exhibition of the French Horticultural Society in 1891 to get ideas for his garden in Giverny, this at a time when new botanical hybridizations were being imported to France from China and elsewhere. Many of Monet's discoveries found their way into his home garden, chief among them tulips, anemones, narcissi, Spanish irises, and large-flowered clematis. For Monet, the plants came first: "It took me time to understand my waterlilies. I had planted them for the pleasure of it; I grew them without ever thinking of painting them."

Colorists paint a garden picture by mixing and melding color, texture, and form, all with precise timing. This is, in my opinion, the most daunting type of gardening. Think of perennial borders—what the English term herbaceous beds—and their often daring color and plant combinations. In my mind, perennials are a bit like actors in a play—some are stars, others are background performers, yet all work in unison, with some coming off the stage just as others go on so they always hold the audience's interest. Most English gardeners are colorists, as exemplified by Christopher Lloyd. His property Great Dixter in East Sussex is a must-see on any gardening tour. "The best gardening, as I see it," Lloyd wrote, "achieves a tapestry of plants. All the units touch or intermingle and no canvas shows through.

Others will have a different viewpoint. They will define good gardening as the cultivation of 'perfect' blooms." Gertrude Jekyll is another great twentieth-century British horticulturist in the colorist tradition. If the chapter title ("Plants Do Not a Garden Make") in one of her books doesn't make the point, her text certainly does. She writes, "Having got the plants, the great thing is to use them with careful selection and definite intention. Merely having them, or having them planted unassorted in garden spaces, is only like having a box of paints from the best colourman, or, to go one step further, it is like having portions of these paints set out upon a palette."

One of the many black-and-white drawings Josef Čapek created for his brother Karel Čapek's influential The Gardener's Year, 1929. *Josef's fun and charming illustrations add a wonderful dimension to this seminal book.*

Claude Monet, The Artist's Garden at Giverny, *1900.*

Collectors are dedicated to the hunt. Karel Čapek describes the kind of fever-pitched passion that makes the collector behave like a maniac, raising everything from, in his words, "Acaena to Zauschneria." He writes, "Let no one think that real gardening is a bucolic and meditative occupation. It is an insatiable passion, like everything else to which a man gives his heart." Frank Kingdon-Ward, a British plant collector and intrepid explorer from the early twentieth century, went on adventurous and dangerous expeditions across Asia in search of exotic plants. He is best known for his discovery of the Himalayan blue poppy, still highly rare and coveted over a century later. Most gardeners are in debt at one time or another to plant hunters like Kingdon-Ward who've risked health or wealth, sometimes both, journeying to remote locations to bring home plant discoveries.

Dirt gardeners—those who toil in the soil—savor the backbreaking work of being in the garden, finding a sense of purpose and utility in the work itself. Čapek believes "that a real gardener is not a man who cultivates flowers; he is a man who cultivates the soil. He is a creature who digs himself into the earth, and leaves the sight of what is on it to us gaping good-for-nothings. He lives buried in the ground." I decided being on the ground and digging in it would be the best way for me to feel connected to nature and to understand plants. And so, while I may be a scene setter by nature, it's the dirt gardener who inspires me and whom I try most to emulate. At the end of the day, while there are as many different types of gardeners as there are gardens, there's only one real designer and that's nature itself. Beverley Nichols writes, "The gardener can provide the frame, set up his easel, and sketch the pattern, but as time marches on he must constantly step aside and hand over his brush to Nature."

GREEN THUMBS AND HANDS-ON MENTORS

I grew up thinking that I had no ability to work with plants. In grade school, we were given a potted violet to grow: Mine died overnight and, yet, when I'd visited a friend's house weeks later, hers was thriving. The idea that certain people are born with a natural disposition to make all things grow is one I've now decided isn't true. Henry Mitchell, among others, would back me up. He feels that "there are no green thumbs or black thumbs. There are only gardeners and non-gardeners." Eleanor Perényi would agree, writing, "People who blame their failures on 'not having a green thumb' (and they are legion) usually haven't done their homework. There is of course no such thing as a green thumb. . . . One acquires the necessary skills and knowledge to do it successfully, or one doesn't."

Perhaps this belief echoes my original idea of wanting to be a dirt gardener—anything can be accomplished with a combination of hard work and perseverance. You'll come by your green thumb through the work of gardening. While some people may have a greater instinctual understanding of plants than others, I'm the perfect example that gardening can be taught. My garden is evidence that plants do grow and thrive for everyone, especially when you're lucky enough, as I have been, to have hands-on gardening mentors, mentors who show you how it's done.

My hands-on mentors are Katherine Schiavone and Gaye Parise. A talented landscape designer and gardener, Katherine has been so very generous, helping on every aspect of my garden—from its overall design to planting techniques and care. Katherine introduced me to Gaye, a self-taught gardener. While Gaye appreciates flowers, her heart lies with vegetables, about which she seems to know everything. I've learned that, if you want to become a gardener, you must be around people who garden. You should watch them and do the work they do.

Beyond their professional expertise, I'm also fortunate to consider Katherine and Gaye my friends. Our relationship, like those between many gardeners, is bound by our mutual love of the garden and nature, and of being outdoors. In this respect, the garden not only gives the gift of friendship to the gardener, but also sustains it. Elizabeth Lawrence was right when she wrote that "no one can garden alone." In the garden, everyone—young and old, mentor and novice—becomes a friend.

CHOCOLATE EARTH

If you have good dirt, you'll have good plants. If you don't, you won't. Just as a house needs a strong foundation, so does a plant.

Beverley Nichols wrote, "'Know thy soil' is as vital an injunction to the gardener as 'Know thyself' is to the philosopher." Soil, according to Bridget Elworthy and Henrietta Courtauld, the British gardening duo known as The Land Gardeners, is the "stomach of the plant." They note that there are more micro-organisms in a teaspoon of soil than there are people in the world. They write, "If our soils thrive, our fruit and vegetables will take up more nutrients and we, in turn, will be healthier for eating them." So important are the values of healthy soil that Karel Capek made the connection between it and selfless love, saying "you must give more to the soil than you take away." In a garden, I've come to believe you'll get far more than you give in the long run. Yet I learned early on that to cultivate roots, I'd need to give more to the soil than what I'd take from it.

When I examined the soil, I realized that in restoring it to health I had a big job ahead of me. It was in bad shape. Dry, chalky, and rock hard, it was so compact that not even a rototiller could break through. I learned that I could send soil samples to a university lab and, for a small fee, have them analyzed. Two weeks after I sent off the samples, the report arrived. While it was hard to understand the statistics, the most important message was clear: The pH was too high. People often correlate pH to being sweet (alkaline) or sour (acidic), but like food, a balance of both is best for most people. Layering the soil with compost helped nurse it back to health until it became a pure, smooth yet texturally dense carpet of brown bordering on black, something Eleanor Perényi called "devil's food cake." Truly, as people say, chocolate earth. In fact, it was so beautiful, I didn't want to disturb it with plants; I wanted to get my hands in it and dig.

If soil is chocolate earth, compost is black gold. Manure is the glue that holds it all together. Early settlers let their livestock roam free, which meant that the manure was often miles away from where it was most beneficial—in the farms.

Recognizing this disconnect, George Washington built a "stercorary," which, at the time, was quite innovative. Basically a covered pit, the stercorary was used to store, age, and mix manure. He believed that "the profit of every Farm is greater, or less in proportion to the quantity of manure which is made thereon."

One hundred years later, Charles Dudley Warner wrote, "The love of dirt is among the earliest of passions, as it is the latest. Mud-pies gratify one of our first and best instincts. So long as we are dirty, we are pure. Fondness for the ground comes back to a man after he has run the round of pleasure and business, eaten dirt, and sown wild-oats, drifted about the world, and taken the wind of all its moods." In reading this, another childhood memory came floating back: I had loved digging for earthworms and making houses for them in old mayonnaise jars with holes poked into the lids for air and folded paper for chairs, table, and a bed. Perhaps I wasn't such a foreigner to dirt gardening after all.

Katherine made an observation that stuck with me. She said, "Gardens give adults permission to get dirty as they did as children." And, indeed, I found myself in the "latest" stage of my life rediscovering one of my earliest pleasures. When I'm deep into the dirt, I've often thought of Nichols's words: "To dig one's own spade into one's own earth! Has life anything better to offer than this?"

THE HEALING POWER OF DIGGING

Long before her gardening book *Four Hedges* was published in 1935, the English-American artist, writer, and illustrator Clare Leighton was known for her fine wood engravings, several of which are used as illustrations in her book. On its surface, *Four Hedges* chronicles Leighton's emerging life as a gardener but it's also about so much more. It was in the pages of this slim book that I first stumbled across a sentiment that gave me pause. It was, of course, something I'd already known intellectually but, until Don's death in December of 2019, hadn't yet fully grasped: "There is great healing power in digging. This is so much the case that one is tempted to wonder if any actual electrical power comes up to one from the earth. . . . At any rate, however sulking and rebellious one may be at the start, sensitiveness creeps up the fork into hands and body and legs. Finally the brain surrenders and one is again at peace with the garden."

In late winter 2020, only a few months after Don's death, I went to Connecticut to check on the garden. It was the first time I'd been there since the funeral, and it felt appropriate somehow that I should see it in this barren, raw state. Leighton's writing motivated me to go into the garden and move the dirt around, to see if what she said could come true, even if just a little bit. It was late afternoon, the light low in the sky, the air chilly. As it had been an especially warm winter, the ground never froze. I sat on the grass, pushed the trowel into the soil and it sunk in easily, seemingly just as needy for work to begin as I. I turned the chunks of soil a few times. It still looked healthy with compost, and I soon detected the unmistakable scent of earth. It gave a kind of warmth to the air even though the temperature itself didn't change. I puttered around until my hands were well into the soil, a satisfying place to be. That night I slept in a house that wasn't yet a home, a place that felt strangely foreign to me, especially without Don. Yet when I looked outside the window and saw the rows of fallow earth waiting for spring's arrival, I felt my roots taking form even though the very root system of my life—my relationship with Don—was gone. Leighton's words reverberated in my thoughts, and I imagined the time when I could allow myself to once again enjoy the first daffodils and the freshness of the spring rain pushing away a winter that stayed too long.

It is a fact that, for as much life that a gardener brings into this world, she will also be witness to much death. Accidents happen. Gardeners make mistakes all the time. I certainly have. But these lessons are also a reminder of life's fragility. The changing seasons, with their dewy spring mornings and blankets of snow, tell us that life is fleeting. Everything is temporary and will go to rot soon enough. As Alfred Austin, the poet laureate of Britain from 1896 to 1913, once wrote, "We come from the earth, we return to the earth, and in between we garden."

III. BUILDING MY GARDEN

A good bone structure must come first, with an intelligent use of evergreen plants so that the garden is always clothed. . . . Flowers are an added delight, but a good garden is the garden you enjoy looking at even in the depths of winter.

MARGERY FISH

BONE STRUCTURE AND BOUNDARIES

Once I started planning my garden, I began to think about it in architectural terms similar to those of a house. The soil was the foundation. The fencing created its walls and helped determine the garden's size. The garden beds were the rooms. Its paths were the halls and the flowers and vegetables the furniture. Clare Leighton followed a similar principle, writing that just as "our bodies have skeletons so should our garden have bony structure. It is only upon a firm foundation that the irregular growth of plant and tree can best clothe and deck the garden." If I got the structure of my garden right, I could experiment all I wanted.

In considering the boundaries of my garden, I knew I needed to clearly define the space. The first and most significant decision I made was to combine the previous owner's narrow cutting bed with space I'd chosen for my garden.

I'd initially thought of leaving the cutting garden in place, but doing nothing still meant doing something: I would have had to repair its broken-down fencing and rusted-metal edges as well as redefine the ramshackle paths that had lost their shapes long ago. Perched on a narrow precipice likely created from landfill, the cutting garden always felt as if it was about to roll down the hill—certainly not in keeping with the idea of establishing roots.

Once the excavation work was finished, the space was transformed—I now had a flat, open plateau on which to work. This was exciting. I suddenly had my own space and no longer had to fit into the previous owner's boundaries. While the open space gave obvious guidelines as to the garden's size of forty-eight by fifty-four feet, it also begged for fencing. Creating a garden is a bit like doing a jigsaw puzzle; I always feel compelled to complete the outside frame of the puzzle before tackling the rest. The same applied here: I knew I needed a border before I could think about the beds.

Gardeners often talk and write about boundaries, borders, and structure: their necessity, how to best construct them, and what they reveal about the gardener's personality. Put simply, a garden's boundary refers to its dimensions; a border, the type of material used to define its perimeter as well as the inner beds. Structure refers to the bones of a garden, the elements that give the space its shape; structure provides focal points where the eye can rest so the garden doesn't vanish into the landscape. My garden, as an example, has an underlying order to it, and I rely on orderliness as a way of living. It's also fairly structured. I hadn't realized how much I appreciate symmetry and structure until I tried to organize lettuce. I wanted to extend this sense of orderliness to the central beds by keeping them the same, but nature intervened: One row of cabbage took off; another wilted.

At the same time, however, I love gardens that are loose, natural, and a bit wild as ours is in Long Island. My little experience tells me it may be best to be orderly yet flexible so that something unexpected, inspired—and wholly natural—can also take form. Every gardener knows that the moment their back is turned and the garden is left untended, nature will take over. Penelope Lively writes about how the garden itself cares little for the self-imposed boundaries we gardeners assign it: "Any garden is a defined area, within which the gardener attempts to impose order. The garden resists, and defies the imposition by throwing up docks, couch grass, bindweed and anything else that occurs to it at any spot the gardener has not been watching intently. . . . It does not observe its own boundaries." The euphorbia, she notes, will hop through the hedge to join rank with the red campion, the stitchwort, the primroses. In her words, the garden escapes.

In thinking about what kind of border I wanted, I first considered making a walled garden, which would have been in keeping with the lovely old stone walls that wind through the Connecticut countryside. A stone wall would have also shielded the garden from a view of the driveway only a few feet away. The walled gardens I found most appealing were generally in England and had a

To create the open plateau for my garden, I merged two connecting spaces that share the large hedge pictured here. Tearing up the basketball court and removing the fence (TOP, FOREGROUND) brought that area and a defunct garden (BOTTOM) on the other side together.

lot of breathing room around their perimeter, something I didn't have. A dense stone structure so close to the house would also have felt fortress-like, and it would have been a much bigger undertaking than I could handle.

I studied fences intensely. Because America's gardening history is important to me, I narrowed my research to colonial gardens, many of which used the Williamsburg picket fence. While there have been variations on this style over the years, it didn't seem like the right fit for our somewhat more rural environment. I discovered and fell in love with one that belonged to Bunny Mellon, a much-admired amateur American gardener. Her fences stretched for acres on end at her Oak Spring farm in Virginia. Working with my builder, I decided to try an interpretation of her design, something that would evoke the rugged openness of early America yet would still suit the garden's gentle country setting. I'll never forget when the first side went up. My heart sank: It looked like a corral fence from the Wild West.

But even had the fence been as I'd hoped, part of the problem was context: Mellon's fence ran through open fields, while mine outlined a dramatically smaller space. Poor judgment on my part, but as Elizabeth von Arnim wrote, the "only way of learning is by making mistakes." It was back to the drawing board—literally. Katherine discovered photos of a combination of trellis and fence at George Washington's Mount Vernon kitchen garden. Attractive and simple-looking, the fence followed in the gardening style and tradition of early America, but it was more trellis than fence. I found a photograph of an old Edwardian fence with four rails instead of the traditional three, and with different-sized spacing in between. We tossed around these ideas along with several others, which Katherine drew in her typically precise and clear manner, reflecting her skills as a trained architect. After the eighteenth or so revision, we came up with something we both liked.

To land on the final design, we considered the number of posts and the distance between them (they vary on the longer versus shorter sides), rail height and thickness, and the type of wood. In the end, I settled on white oak, which is

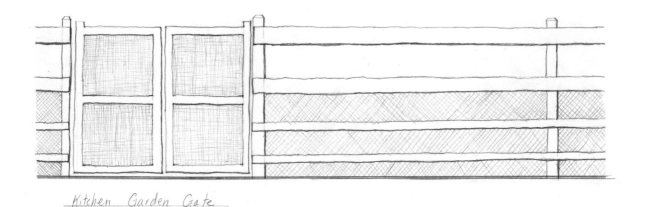

Kitchen Garden Gate

supposed to weather to a lovely gray. To keep out small animals, we opted to dig old-fashioned chicken wire eight inches into the ground. I wanted the gate to look like a simple screen door, complete with hinges—more decisions.

As soon as the posts and rails went in, we knew the fence would be just right. It looked solid, crisp. The wood had a lovely, fresh warmth to it. We were proud and perhaps appreciated it even more because of our extra work. All in all, it was a straightforward fence, very much like the garden itself, yet I think the subtle details, which most people don't notice, make all the difference. But that wasn't the end of the saga.

By midwinter, the rails were so black at points that they looked as if the wood had been singed in a fire. It turned out that sap leaking from the oak had reacted with the nails, which were made of the wrong type of metal, and blackened the wood. Winston Churchill wrote, "Every garden presents innumerable fascinating problems"; this was a fascinating problem I hadn't envisioned. Then I reminded myself of American quilters, some of whom were known for inserting a sewing mistake into their quilts as a reminder that no one is perfect.

ABOVE: *Katherine Schiavone's architectural drawing of the new garden fence.*

There's a certain kind of freedom in making mistakes in the privacy of one's own garden—no one's watching. The stakes, so to speak, are low. Von Arnim wrote, "Last night after dinner, when we were in the garden, I said, 'I want to be alone for a whole summer, and get to the very dregs of life.' . . . Nobody shall be invited to stay with me. . . . I shall spend months in the garden, and on the plain, and in the forests. I shall watch the things that happen in my garden, and see where I have made mistakes." In today's fast-paced lifestyle, to be able to slow down and make mistakes here and there while trying out something new is precious. As it turned out, I needn't have worried too much about the fencing: After much thought, we decided to paint it. I've been very happy with its dark green color; it's an ideal backdrop for the flowers.

Robert Pogue Harrison wrote, "A garden is literally defined by its boundaries. . . .they keep the garden intrinsically related to the world that they keep at a certain remove." It also tested me on the very character traits I'd heard gardeners possess: patience, perseverance, and inner calm. All told, it took nine months to design, construct, and finalize the fence. I'd never before given a second thought to fences but over the course of that time I came to see them as lessons in personality, proportion, and, yes, more than a little patience.

When I ran across Harrison's quote some months into building the fence, I felt better about the intensity with which Katherine and I had worked. Harrison was right: The fencing, rather than closing me off from the world, instead opened me up to the wonders inside its borders. Standing inside the garden, the rest of the world—its daily pressures and chaos—fell into relief. When summing up the qualities of a garden, English poet and textile designer William Morris wrote, "Large or small, it should look both orderly and rich. It should be well fenced from the outside world. It should by no means imitate either the willfulness or wildness of Nature, but should look like a thing never to be seen except near a house." Now when I look at roadside fences, I think of how silly I was; I can't reinvent the wheel. Or better yet, I should have listened to Jefferson, who was fixated on organic fencing. He purchased four thousand hawthorn trees to create a wall of hedges at Monticello.

Katherine's watercolor of my garden. A Christmas present to me in 2020, it captures the garden's spirit in a timeless way and is now one of my favorite paintings.

NEW ACQUAINTANCES IN THE FAUNA WORLD

We completed our home renovation in early 2019. After enjoying a celebratory lunch with my team, we inspected the finished house and spotted a large, brown owl sound asleep in his nest on the wall outside our front door. He looked peaceful and completely undisturbed as we took photo after photo. Unbelievably, no one had seen an owl in the wild despite there being more than one nature lover in the group. We couldn't restrain our excitement and awe, and certainly made more noise than we should have, yet the owl didn't move. The next morning, the first thing I did was to check on him and I wasn't surprised to see that he'd gone. I decided that his visit was a good omen about our lives to come. It was the first sign for me that the house could become a home—a place to nest. At the very least, his appearance confirmed that we now lived in the country despite our proximity to New York City. A year later, that message hit home in a much more ominous way.

The cold spring days of the pandemic lockdown were dark and felt never-ending. It seemed the only color to be found was in our tulip garden. During a particularly long stretch of dreary weather, I was awakened suddenly from my sleep in the depth of a chilly April night by the sound of something—or someone—screeching, the likes of which I'd never experienced before. I was certain I was having a nightmare. The crazy screeching continued, echoing through the woods until it seemed to land, reverberating, right outside my window. It took a few seconds for me to realize I wasn't dreaming and a minute or two more to convince myself that I wasn't stuck in a horror film or a gruesome fairy tale, a coven of witches circling nearby torturing their victim or being tortured themselves or worse.

I ran upstairs to wake my children. By the time I'd reached them, the screeching had stopped. For the rest of that fitful night I slept with our dog in the guest room, which was closer to their rooms than my own. I learned the next day that the screeching was so loud that our neighbors also heard it, and guessed it was, in fact, a pack of ten or twelve coyotes chasing a deer. Nonetheless, for the next month or so, I slept in the guest room until I finally forced myself to go back to my own bed. I had to get used to the nighttime sounds of the country.

Fast-forward to a balmy summer evening in the garden when I'd brought some snacks outside for a friend and me to enjoy. As we were eating, a big, black crow swooped down and stole an entire piece of cake. Just a few days later, the same crow arrived once more, this time making off with a whole slice of pizza. Herb, our scarecrow, stood by, oblivious and of no help whatsoever. Was his friendly face not intimidating enough to fool the crow? Are scarecrows old wives' tales? Who knows? The crow has come again, but I've learned my lesson about leaving out food.

BORROWED LANDSCAPE, BORROWED IDEAS

Douglas Chambers wrote, "A true garden is never apart from its landscape." I thought of this when I first stumbled across the term "borrowed landscape" while reading up on Capability Brown, the eighteenth-century landscape architect commonly known as "England's greatest gardener." The term means extending one's view by taking advantage of the vistas offered by the neighboring properties. If I had created a walled-in garden, I would have closed myself off from seeing the rolling hills of our neighbor's property. This borrowed landscape, coupled with the expansiveness of the sky above, gives me a sense of openness and enhances the feeling I'd longed for of being in the country.

The concept can also apply to borrowed ideas: Gardeners are forever borrowing ideas from one another; copying is how you learn. Even the most legendary landscape designers copy. Everything is a bit "fair use" because even when you borrow something exactly, it's never quite the same—conditions vary from garden to garden. There are too many variables to make an exact copy, which is why, perhaps, gardeners have so much leeway. Russell Page confessed, "I know I cannot make anything new. To make a garden is to organise all the elements present and add fresh ones, but first of all, I must absorb as best I can all that I see, the sky and the skyline, the soil, the colour of the grass and the shape and nature of the trees."

Happily and not surprisingly, most gardeners are innately generous, and are usually delighted to share their plants and their ideas. I have come to certain viewpoints on my own only to later read about them from a guru. It was sometimes frustrating when this happened; however, it further confirmed the existence of universal gardening truths. Or simply opinions! Roy Strong, for one, had a strong opinion on the generosity of sharing plants, writing, "Plant gifts, as all gardeners know, produce agony as well as ecstasy. Ecstasy at the generosity, agony over whatever to do with it."

My favorite gifts are the simplest: homemade syrup from a friend's maple tree upstate, fresh eggs from another friend's coop, a holiday wreath made from pine cones and dried flowers. I've always wanted to be able to give gifts like this of my own. In fact, it was one of the appealing aspects of having a vegetable garden. Now I've had that chance. It's fun to go into the garden as friends are leaving and together cut lettuces, kale, and basil; pull radishes and whatever else is ready; and fill a brown bag with produce for their evening salads. There's a lot of joy in growing vegetables and flowers, and even more in sharing them with others.

IV. CONTENT IS EVERYTHING

Give me a garden full of strong, healthy creatures, able to stand roughness and cold without dismally giving in and dying.

ELIZABETH VON ARNIM

DESIGNING THE BEDS

While Katherine and I were still designing our first try at the fence, I began to think more concretely about the actual garden, the layout of the beds, and the plants that would go in them.

In my dreams, I imagined a typical French vegetable garden, with its many beds of different shapes, each of one vegetable with some flowers mixed in, next to an ancient château or farmhouse under a late fall sky. This type of grand French garden reflects the French love of food, terroir (an appealing word that describes a combination of soil, climate, and sunlight), and order. Their kitchen gardens are known as potagers (the French word for "soup" is potage) and are equally about beauty, form, and function.

Once the fence was in place, it was fun to stake out possible ideas for the beds. In the past, whenever I saw a garden's plan in a book, I'd quickly turn the page. Now, these are the pages I study the most. I'm fascinated by the decisions gardeners make: how one bed relates to another, their shapes, the distance between them, their design.

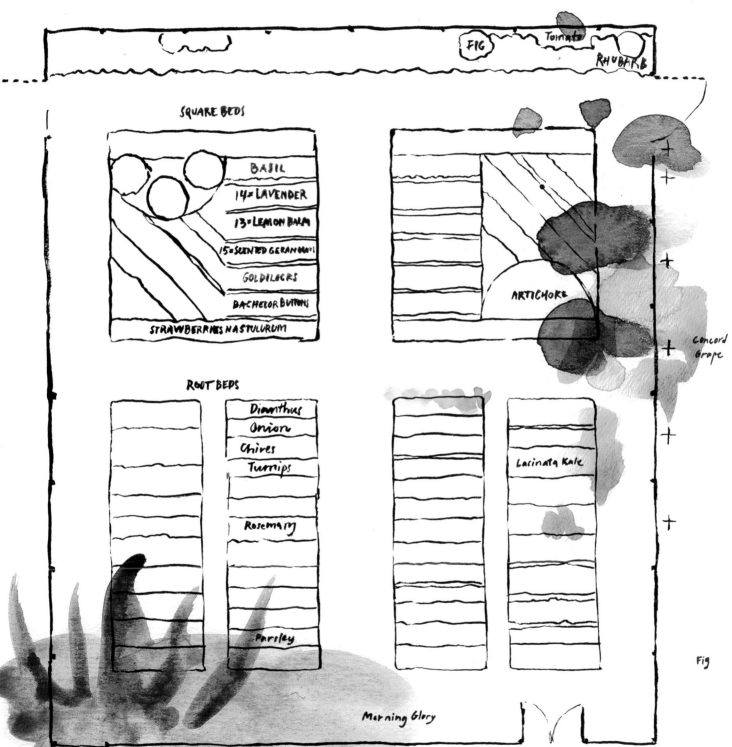

FIG   Tomato

RHUBARB

SQUARE BEDS

BASIL

14 × LAVENDER

13 × LEMON BALM

15 × SCENTED GERANIUM

GOLDILOCKS

BACHELOR BUTTONS

STRAWBERRIES NASTURTIUM

ARTICHOKE

Concord Grape

ROOT BEDS

Dianthus

Onion

Chives

Turnips

Rosemary

Lacinata Kale

Parsley

Fig

Morning Glory

Himrod Grape

Ficus Folium

Fernando Caruncho's garden layouts particularly appealed to me. The driving force behind his garden designs is Greek geometry and the grid, which he softens with trees and shrubs. I quickly settled on a grid of rectangles and squares for my garden, the squares being the same width as the two rectangles together. I first designed side beds, but then decided the space would be too crowded. Instead, I put in climbing plants on both sides of the fence, which I especially love. Depending on the month, blackberries, morning glories, grapes, and roses share space with sweet peas, currants, hollyhocks, and—because Gaye couldn't imagine a garden without them—tomatoes. I'm glad I followed her advice to plant them. There's nothing better than eating a sun-warmed tomato fresh off the vine. It reminds me of childhood days when I'd asked my mother for a tiny Morton's salt-shaker, the perfect size for one tomato. I kept it in my pocket, plucked tomatoes from our patch, and salted and ate them whole.

Paths were a quandary, their width in particular. I followed Perényi's advice: "Have it broad enough to fit yourself and a friend, as attractive as possible to the eye and kind to the foot." To measure my own paths, I asked friends to walk them with me, and then gauged how we best "fit" with one another. I settled on two different widths: six feet and four feet, both of which are wide enough to allow easy access for people, wheelbarrows, and a lawn mower.

While the grid was great for layout purposes, I still had to decide what to plant in the garden beds. When Russell Page dreamed about his own future garden, he planned that "each bed will be autonomous, its own small world in which plants will grow to teach me more about their aesthetic possibilities and their cultural likes and dislikes." This, I felt, is where the heart of gardening begins. I began to plot out which plants would go where and wondered what they would teach me.

James Fenton summed up the key to a garden: "The important question is what do you want to grow? . . . Content is everything." Mary Keen, one of England's most esteemed garden designers, takes it one step further: "Knowing how to grow things is important if you want to make a garden, but not as important as people make out; it's knowing where to put them that matters."

The spirit of my garden plan was clear; the challenge, of course, was how to achieve it through plants. I knew from the start that I wanted to walk into rows of lush lettuces and abundant peas. Moving from there, we knew it made sense to keep almost all the vegetables in the rectangular beds, and the flowers and the herbs mostly in the square beds.

I met with Katherine and Gaye several times over the winter for their advice and to hone my choices. While attempting to have plants I thought I'd love to look at and vegetables my family loves to eat, I've been game to try new ones. For instance, Gaye strongly recommended white turnips (a vegetable I wasn't sure I'd ever eaten), and a puree of them is now almost as delicious as fresh peas to me.

We tried to grow a lot from seed—their packets contain helpful information on everything from days to germination and maturation to soil recommendations and planting depths. To get a head start, we grew some produce by a sunny window indoors, and Gaye had a friend who raised several flats for me in her greenhouse. Almost all vegetables are annuals, and while there are some "cut and come again" lettuces, most vegetables bloom, are harvested, and then the dirt's ready for the next seed or seedling. I didn't know that when I first started my garden, but this is why kitchen gardens are often more work than other types of gardens. They need more constant tending, harvesting, and replanting.

Scientifically, it's smart to plant certain plants together. Just like people, plants naturally get along with some better than others, so we put root vegetables—carrots, radishes, and turnips—in the center two rectangles and brassicas—cauliflower, broccoli, and cabbages—in the outside beds. I also included certain plants for pollinators like lavender and bachelor buttons, which bring the bees and make the surrounding plants more profusive, yielding more flowers and bigger vegetables, and "beneficials," such as chives and rosemary, which repel the insects you don't want. Beneficials can also improve the taste of vegetables; for instance, basil planted next to tomatoes will improve their flavor (curiously, in salads, too). The middle beds are virtually the same; Katherine calls them twins and the outside ones sisters, as there's more variation.

Ellsworth Kelly, Sunflower, 1983. Of all of Kelly's pencil drawings, this is a favorite.

It's essential to pay attention to the sun. We planted on the preferred North–South axis to get the most sunlight. Still, one side of the garden gets much more than the other, which dictated which plants went where. Lettuces need some shade, as the hot summer heat is too much for them. On the other hand, lavender adores the sun, and artichokes grow quickly in the warm summer heat. A small break in the trees gave just enough sunlight for the sunflowers to shine.

My dream of lush lettuces to greet me when I opened the garden gates came from photos of French potagers. The design of the square, flower-dominated beds, with their mix of diagonal and horizontal rows, came from studying garden plans, specifically Bunny Mellon's plans for her garden in Virginia. Inspiration for sunflowers came from Gravetye Manor, the former home of landscape designer William Robinson. They were never before a favorite of mine, but they looked so robust, so happy, reaching high above the tall brick wall on the late September afternoon when I visited there that I thought they'd be ideal on my fence.

I've loved the idea of old-fashioned hollyhocks, morning glories, and bachelor buttons, mainly because of my childhood storybooks, like *The Bumper Book*, where they seem to grow freely, so I planted those along the fence. My Italian blood made me long for grapes and figs. So I planted grapes on the fence and, as fig trees need to come inside for the winter, put them in big wine barrels and placed them in the beds. French fields of exuberant lavender are stunning, and I pictured loose, fragrant rows, but they never took off. Daisies and flowering apple trees always make me feel as if I'm in the country, and I thought they'd be great together. Unfortunately, they don't bloom at the same time so that idea had to go. Instead, I have sweet peas, currants, and roses. It's been nice to see the currants mix with the roses in a pretty way.

Again and again, I've read or heard about the glory of spring tulips and thought about how lovely it would be to herald the season with colorful, vibrant blooms. As most vegetable seeds aren't put into the ground until after all frosts, I had time and space to do so. I looked through catalogs to choose those I like the most, parrot tulips having always been my favorite. Katherine wisely suggested that we have early-, mid-, and late-blooming tulips so we'd always have something wonderful to look at. I never guessed how much I'd love them. I also wouldn't have guessed how many we'd need. We planted almost a thousand! Instead of creating a formal design, we mixed them up, each stage of bloom having its own brown bag, and we had fun, just pulling them out and planting, so that whatever came up would be purely haphazard. I was happily surprised by their arrangement, but I think any combination would have worked. They are joyous and so plentiful that I could cut many to bring inside and not disturb the garden's look.

Deputy Bob

PLANTS AND TREES OF NOTE

While vegetables and herbs are the backbone of my garden, flowers are equally important in their own way. I'd feel something would be missing if it was only green. Many plants—and certainly some trees—capture my imagination and bring back memories. Here's a sampling of some:

Dahlias

One of my most memorable luxuries happened several summers running after we'd moved into our house in Long Island. A lovely, kind man named Tony DiPace raised dahlias, and every Friday afternoon from about mid-August to Labor Day, he would drop off a few big buckets of inexpensive blooms in all their rich, deep, late summer colors. These deliveries brought end-of-week joy. They were also a bittersweet reminder that I'd soon resume the normal pace of urban life, returning as we would to the city for the fall.

At the time, dahlias weren't popular. In fact, when asked which flower was my favorite, I felt rebellious to say dahlias! For some, dahlias have never gone out of style. Henry Mitchell was a fan, writing, "Sometime in late August there suddenly comes a hint—maybe you feel something in the air—that summer is passing. It is then that dahlias are in their glory, and while none have yet been bred that are quite as large as TV sets or as bright as atom bombs, they will bloom magnificently and conspicuously enough through September and October, when few other things do. . . . I cannot think of a more vigorous, spectacular, up-and-at-'em flower for late summer. Regular tigers they are."

Public opinion has changed dramatically in the last decade or so. Dahlias are now so much the rage that those of us who've loved them for years are suffering from dahlia fatigue. An English landscape designer I know recently said they're so ubiquitous that he no longer likes or uses them. I

wouldn't go that far, but admittedly their appeal has faded for me, too. Still, late summer wouldn't be the same without them. We were going to plant them in one of the square beds, but as they're so dominant, I thought they'd aesthetically drown out the more delicate plants, so we created a raised bed outside the fence specifically for them.

They've come up in small spurts, never offering that huge bolt of color that tulips do nor acting as much of a screen for the driveway as I'd hoped. Because of this, I've been cutting and putting them in vases throughout the house and giving them to friends who stop by.

There's a theory espoused by the Land Gardeners and others that says that when the tulips are done, you take the bulbs out and plant dahlias, and then do the same in the fall after the dahlias have finished their show. This seems logical but it isn't practical for my garden this year; I need the space to grow more vegetables, not flowers. I'm keeping this in mind, though; maybe I'll try it next year.

Kale

Kale has never been one of my favorite vegetables although my daughter, Serena, and her friends love it. While to me, it's had its year or two in the sun as the newest vegetable on trendy restaurant menus, its popularity has seemed to have mellowed, and you can get a Caesar salad made with romaine once more. It's a strange-looking plant, not nearly as attractive as the lettuces planted next to it. The leaves have an almost leathery, bumpy texture, which I suppose is why the lacinato variety is nicknamed "dinosaur" kale. It has such a long stalk that it reminds me of a palm tree. However, for sheer productivity and health benefits, it merits a spot in this list of notable plants. We cut and cut the leaves all summer long—they keep coming up and Serena keeps eating them.

Pumpkins

When our children were young, I attempted several years running to plant pumpkins in our garden on Long Island, thinking they'd be a fun discovery come October. I followed the packet's directions and planted the seeds at the appropriate time, the beginning of June. The pumpkins came up, small and orange, but they arrived at the beginning of August, two months early. I decided to try my hand at them in Connecticut, planting seeds in mid-July. They popped up small and green and in mid-September! Over the next weeks, the green became striped with orange and gradually they became more orange. By late October, they reached their final fully orange color. Although they were small in size and did not grow much bigger, I'm thrilled to have finally managed to produce fall pumpkins. Trial and error is, of course, the right way to garden, my pumpkins being a prime example. I'll persevere and next year I'll see what happens if I plant them right after the Fourth of July.

Rhubarb

During our first summer season at our home in Long Island, a bunch of huge-leafed plants sprouted in their own makeshift patch. I found them obnoxiously large, ugly really. While I didn't know what they were, my friends, both experienced gardeners, did and they couldn't wait to come and dig them up. While I watched them do this in great excitement, I scratched my head. What did they see in them?

As we all know, people change. Years later, I've come to appreciate rhubarb, especially since I love rhubarb pie. I decided to plant one in a corner of the garden. While it hasn't grown to be as large as those I remember in Long Island, I now like its purple stalks and wide, green leaves. As it's a perennial, I look forward to watching it grow and, I hope, become huge.

SCENTED
GERANIUM

Roses

While this is probably an exaggeration, it seems to me that there has been as much written about roses over the course of time as all other flowers combined. Not only are roses the Western world's most popular flower, but they also seem to be its most controversial. I imagine if I put legendary writers and gardeners in a room and throw out the topic of roses, civility will fly out the window.

Perhaps the loudest voice on the subject would be that of Christopher Lloyd, the owner and creator of Great Dixter, one of England's most admired gardens. Having been a world-famous rosarian for fifty years, he suddenly tore them out of his garden beds in 1993 and replaced them with tropical flowers and dahlias. Jamaica Kincaid reminisced: "Just as I was mulching and covering up the roses for winter, I came upon Christopher Lloyd saying in the increasingly beautiful-to-look-at *Gardens Illustrated*: 'I got fed up with all the troubles roses bring in their train. They get a lot of diseases and you can't replace a weak bush without changing the soil. They're quite disagreeable and make a very spotty effect even when they're flowering—a whole series of blobs. The climbing ones are quite shapely but the bush roses are pretty ugly.' . . . The capriciousness of a gardener! I see that I shall be overwhelmed by the number of roses I will have ordered by next spring, so that twenty years hence I will firmly denounce the whole idea of growing roses."

You can sense the brimming conflict in James Fenton's reminiscence of his weekend at Lloyd's Great Dixter. About twenty years ago, Fenton wrote articles on his English garden in *The Guardian* but he did so "reluctantly" because, in his words, "I did not wish to encroach on the territory of the great Christopher Lloyd, who was then enjoying his final years of fame as a garden writer, and whose work I had followed devotedly for decades." Invited to visit Lloyd and realizing that

OPPOSITE: *I found this nineteenth-century watercolor by an unknown artist in a small antiques shop in Paris more than twenty years ago. I love the painting so much that I wish I could hang it in almost every room in our house.*

another chance to see the man's garden might not come again, Fenton cleared his schedule to stay the night, wanting to see the garden at dusk and dawn. He writes, "At dinner, I was questioned a little perfunctorily about my own garden, which I knew my distinguished mentor would not like the sound of. For at the heart of it was a rose garden, and Chrissy had recently waged a campaign against rose gardens, having dismantled his own."

Writers of past eras rarely criticized roses. But today, with everyone speaking more freely, they do. Only a few years ago, Charlotte Mendelson said, "Roses are essential to our native idea of a garden. They litter our myth and language. . . . Roses are the flowers to end all flowers, the ones we sing and paint and compare our beloveds to. . . . Who doesn't love a rose? Well, me." She describes two common colors of roses as being "leprosy yellow" and "knicker pink," going on to note that they are "frequently overbred and underwhelming, prissily miniature, or oppressively rampant, with stupid names . . . and delusions of grandeur."

Roy Strong sums up much of the reason why, writing, " 'A rose is a rose is a rose. . .' Well, for how much longer I don't know. . . . Roses are beginner's plants. They grow quickly and provide instant bloom." He does, however, go on to say that he loves their names and the associations they conjure. I'd agree. Strong writes that the names "Lord Penzance, Lady Waterlow, Louise Odier, Docteur Jamain and Ferdinand Pichard fill the garden with old friends. How lucky to live on as a rose."

Michael Pollan sees the rose in a more lighthearted way: "Chrysler Imperial is actually the name of a rose. So is Sunsation. And Broadway (a two-toned wonder gaudy as a showgirl). Hoola Hoop. Patsy Cline. Penthouse. Sweetie Pie. Twinkie. Teeny Bopper. Fergie. Innovation Minijet. Hotline. Ain't Misbehavin'. Sexy Rexy. Givenchy. Graceland. Good Morning America. And Dolly Parton (a rose with, you have probably guessed, exceptionally large blossoms).

OPPOSITE: *Ellsworth Kelly*, Rose, *1983.*

Cy Twombly, Blooming, *2001–2008.*

It seems to me that the world conjured by these roses is precisely the one we come to gardening to escape." He makes me laugh as does my favorite painter, Cy Twombly, whose blunt opinion was brutal: "I hate roses. Don't you? It's all right if you can hide them in a cutting garden. But I think a rose garden is the height of ick." And, yet, he often painted them.

Having now studied the pros and cons of roses, I decided to start slow and planted two ramblers on the fence. On Vita Sackville-West's recommendation, I planted Lawrence Johnson and Madame Alfred Carriere, which in Sackville-West's words is "white, flushed with shell-pink, has the advantage of a sweet, true-rose scent, and will go to the eaves of any reasonably proportioned house." Both are taking a while to settle in; only two small flowers bloomed. While I've learned you can't rush the classics and I'll certainly see what happens next year, I'm on the fence about keeping them in the years to come.

At the end of the day, the voice that echoes most prominently is that of Fernando Caruncho, who said, "For me the flower is the ultimate expression of the beauty of nature in its vegetable sense, so my respect for roses and flowers is the same as that for my dreams, because roses represent dreams."

Sweet Peas

Someone told me years ago to never put fragrant flowers in a table centerpiece as their scents conflict with those of the food. I've kept that in mind, but I haven't adhered to it religiously. Garden roses from our local florist in midsummer win the day. For Roy Strong, sweet peas are problematic, a sentiment shared by the great Victorian plantsman E. A. Bowles, who Strong quotes as saying, "'A dinner-table decorated heavily with Sweet Peas spoils my dinner, as I taste Sweet Peas with every course, and they are horrible as a sauce for fish, whilst they ruin the bouquet of good wine.'" Strong goes on, telling "flower arrangers please take note." While I'll steer clear of sweet peas ruining a meal of a guest highly attuned to scent, I love having them on my desk or next to my bed.

I've tried to grow them with mixed success. A lot of plants are needed to achieve a small bouquet. But they're worth it, for their delicate petals and, of course, their fragrance. I've always thought "sweet pea" is a charming nickname, old-fashioned though it may be, and now wonder why it's one distinctively scented flower that isn't an inspiration for perfume.

Trees

A young man for his bar mitzvah, new parents for the birth of their firstborn, and a wedding couple for their registry asked for trees as gifts. I first came upon this idea twenty years ago reading the work of Carla Carlisle, the writer and devoted gardener, who asked for a tree for her birthday. I've clearly paid attention to every instance since. A garden is temporary at best but a tree lasts. A tree is a statement. It's a lifetime. It makes sense to me that a tree would make a lovely gift, because it gives back both to the land and to all who enjoy it.

As I wanted to create a sense of permanence in Connecticut, I thought it would be a treat for each member of my family to choose their own tree, which we'd plant in various places around the property, thus giving each of us a proprietary interest and pleasure in watching "our" tree grow and flourish. Don chose an American beech, a handsome native tree, which holds its leaves year-round. I hadn't thought much of it at the time, but I know that it's one of the most stately native American trees, yet also common. My husband's simple upbringing and successful life mirror the qualities of the beech perfectly. The American beech is both a shade tree and an ornamental tree, so it gives to others while having a beauty of its own, something that was also so true of Don.

Our son chose a bur oak, a mighty tree of long life. Providing food and shelter to many animals, the tree is generous, just as he is. Highly tolerant, it can adapt to just about anything. Oak trees are tall. William is, too.

Our daughter and I both chose trees largely for their aesthetic beauty. Serena asked for a lilac and we found a tall, leggy one, just like her. The flowers

vare delicate, with a heavenly scent. It's a tree that blooms effusively but still withstands all sorts of weather, a quality that mirrors a similar hardiness in my daughter. I knew exactly what kind of tree I wanted—a crabapple, mainly because my very first memory in life, as I've mentioned earlier, is of a crabapple tree with pink blossoms.

We also decided on a group tree, a horse chestnut. Since marron means "chestnut" in French, we thought it fit nicely in the familial sense and because it worked with the two already planted in front of the house. After planting these trees, I later realized we'd done something Clare Leighton wrote about: "There can be nothing casual about planting trees. After to-day the place will never again look the same. We shall have changed the shape of the landscape."

In addition to our family of trees, we also planted a small orchard of six old, gnarly apple trees, with different varieties of apples. In choosing them at a commercial apple farm, where I wandered through acres and hundreds of trees, we examined them for their beauty, especially for their shapes, and for the kind of apple they produced. We chose two Cortlands, two Idareds, one Winesap, and

Wood engravings by Clare Leighton from her book Four Hedges: A Gardener's Chronicle, *1935.* ABOVE: *Blackbird on Nest;* OPPOSITE: *A Lap of Windfalls.*

one Burgundy. Cortland apples are good for eating soon after picking; they're good for baking, too. The Winesap apples are primarily used for baking and for making cider, something I've yet to do. The Idareds are sweet and mild, and go with anything. While I like the Burgundies for snacking, I chose one for its bright red color. Oddly, the apples on that tree are more yellow than the others, but this is just for the first year. Like me, trees take time to acclimate to a new home.

Hermann Hesse, in his book *Wandering: Notes and Sketches*, sums up a tree's majesty: "For me, trees have always been the most penetrating preachers. I revere them when they live in tribes and families, in forests and groves. And even more I revere them when they stand alone. They are like lonely persons. Not like hermits who have stolen away out of some weakness, but like great, solitary men, like Beethoven and Nietzsche. . . . Nothing is holier, nothing is more exemplary than a beautiful, strong tree." Now when I look out the window and see these trees, I also see my family, the roots of who we are. When the beech tree's golden leaves glow, I always think of my husband, the light he brought to us and his spirit.

Plant Names

Morning glory, hollyhock, and sweet William are in my garden mainly because of their names, which I find full of old-fashioned charm. The question with plant names is whether or not to rely on the common name or to take the plunge and learn the Latin version, a nomenclature and classification system that has been used worldwide since Swedish botanist Carl Linnaeus invented it in the early eighteenth century. Each plant is categorized under a long list of classifications, but with the binomial nomenclature system, plants can be identified using just two of those terms. The genus name comes first and is always capitalized, followed by its species name, which is always lowercase. Combined, they create a binomial name. Using *Alcea rosea* as an example, the genus name refers to the type of flowering plant—*Alcea* or more commonly hollyhock—in the larger mallow family Malvaceae. The species name refers to the specific type of hollyhock.

I've relied on the common names so far, and they've held me in good stead. However, if I become a more advanced, serious gardener, I'll study the Latin system. If and when I do, I'll follow Russell Page's system for learning names. He writes: "I learned about plants rather quickly. By dint of holding them I began to suspect from their 'feel' and their appearance what kind of conditions they would enjoy and soon I began to be able to guess their place of origin. I learned their names simply by writing down in full the name of any plant I saw for the first time. Even now when I see a plant which I cannot name for the second or even for the fiftieth time, I write out the name, in the end one learns it." That's a bit tedious perhaps, and reminiscent of school, but it's certainly practical. Failing to remember the common or the Latin name, I suppose you could make up your own and name your favorite plants after people you love, and the ones that give you trouble after those you decidedly don't. Of course, that won't get you very far at the nursery.

Although I haven't learned to use the Latin names fluently, I appreciate their value. Some Latin is easily translatable, such as the genus *Tulipa* (tulip) and the family Liliaceae (lily) of which the tulip is a member. Were it not for the Latin I wouldn't have made the connection between tulips and other sorts of lilies. Now it seems quite obvious.

After planting almost one thousand tulips, I began to recognize both the similarities and differences between them. It's surprising that there aren't many Latin names for the various species, especially given how popular tulips have been since tulipomania in early seventeenth-century Holland.

Because we planted the tulips with such a carefree attitude—mixing up handfuls of bulbs as we went—they came up as a jumble, with single-flowered variations next to doubles and lily-flowered with the fringed. My favorites are the parrot tulips as well as the double and the fringed varieties, all of which are quite different from the simpler, vase-shaped petals of *Tulipa clusiana*. One of the few named species, *Tulipa clusiana* is native to Asia, Iraq, Iran, Afghanistan, and the Himalayas, and over time naturalized in western Europe; it is the species from which the many varieties we know today have evolved.

Ellsworth Kelly, Tulip, 1984.

The Beauty of Scent

A professional photographer once told me that he wished he could photograph scent. I'm sure many other photographers and artists feel the same. Illusive, ephemeral, mercurial, mysterious, uncatchable yet retrievable—that's the nature of scent. Of the five senses, scent has the most to do with memory—and memory is a powerful factor in determining a garden's essence.

Not only does scent prompt memories, it announces arrivals. How often do we smell something, knowing it's there before we see it: a Thanksgiving turkey in the oven, for instance, or freshly mown grass at a neighbor's house. For each of us, I'm sure many examples come to mind, pleasant and sometimes otherwise. One early evening this past spring, Gaye and I were digging in the garden when I detected a gentle, delicately sweet fragrance. She thought it must be the wisteria; I guessed it was the lilac around the corner. We sniffed all the plants and couldn't put our finger on it and, as if to prove that scent is elusive, it disappeared almost as quickly as it came, perhaps blown in and away by the wind? As lilac is my favorite scent, our Connecticut trees brought back memories of my childhood and the old, very fragrant bush planted near my bedroom window. Fast-forward twenty years to my discovery of another equally old one on our back steps in Long Island, which cheerfully bloomed for almost twenty years. And now we're blessed with three similar ones as well as my daughter's choice of tree. Coincidentally, her bedroom is right above them and I sometimes imagine how lovely it must be to wake up to their fragrance.

I've come to realize that since a flower's scent is an integral part of its being, I should spend some time studying it, and have determined that for me this is best done by closing my eyes. Another way is to walk around your garden at night, as scents are stronger then. Eleanor Perényi said that, "Everyone knows that, but not that they are also different." Until reading this in her book, I didn't. From Isabel Bannerman I learned that white flowers are often the most highly scented, as they're frequently pollinated by night-flying insects.

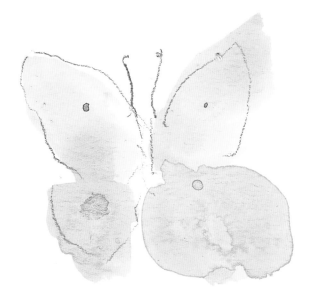

Bannerman writes that "Sight, sound, and touch are physical senses but smell and taste are chemical; in the exchange in some extraordinary way what we smell becomes part of us. . . . the olfactory membrane is the only place in the human body where the central nervous system comes into direct contact with the outside world." Perhaps this is the allure of oils and perfumes. Scent is the essence of a thing, its very nature, wholly unique to the person, flower, or herb. Its allure is alchemic. And yet we often struggle to find the right words to describe a scent much in the same way true beauty stunts our vocabulary.

People debate whether gardens are an art. Whether they are or not, they certainly are an impermanent one. But so is beauty as is scent. Think of lily-of-the-valley, which stays in bloom in the ground for only a short time, in a glass for only a day or two. They're fragile. Here one step and gone the next. Yet that beauty—and its scent—stay with you.

LEMON VERBENA

CHAMOMILE

BASIL

SAGE

LAVENDER

LEMON
BALM

SCENTED GERANIUMS

VALERIAN

ANGELICA

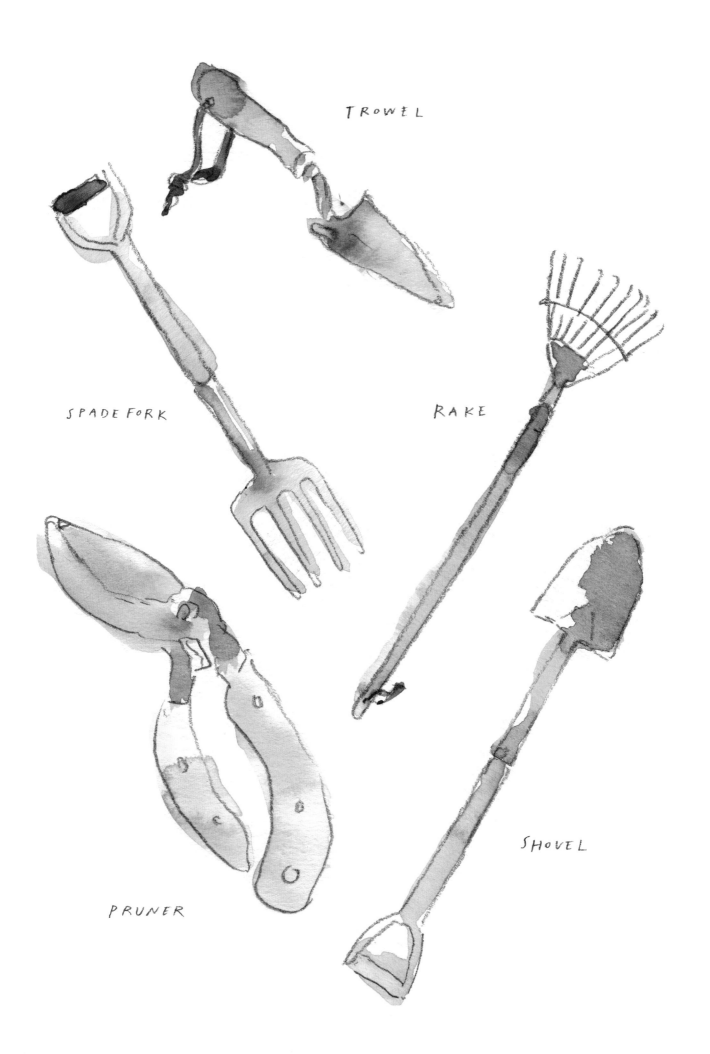

TROWEL

SPADE FORK

RAKE

PRUNER

SHOVEL

TOOLS

Most gardening tools are so simple and straightforward that it's easy to understand why they've been used for centuries. I find it comforting that modern-day technology hasn't improved much on their original designs. Early on in my gardening project, I read that it's wise to invest in the most expensive tools you can afford, and that's good advice. You want a strong, well-made tool when you're working in the garden, something that can last a while and put up with hard labor and all sorts of weather. Today, many tools are handcrafted with care and built to last a lifetime. Like a golf club or a tennis racquet, a shovel or a trowel will feel more and more like yours the longer you use it. Even better, as your gardening tools age, they develop a lovely, well-worn look, something I will appreciate with hard-earned pride.

As this project was moving along, Don surprised me with a wonderful Mother's Day present: a practical gardening bag filled with several hand tools, including what's become my favorite, a trowel, which he'd hunted down. It had been hand-forged, created to last. At the time, it was one of my favorite gifts I'd ever received, and all the more so now. Whenever I use any of the tools from that bag, I think of him.

Gardening tools are not only essential, but an extension of your body—sometimes literally. You clasp a spade with your hand and, in doing so, your arm is elongated. A rake moves to the swaying of your hips and arms. A hoe chops as you chop, mimicking your hand motion. Gertrude Jekyll writes that they can even impart a loveable quality: "One feels so kindly to the thing that enables the hand to obey the brain. Moreover, one feels a good deal of respect for it; without it brain and hand would be helpless."

For as many jobs as there are in the garden, there's a tool to help you do it, and for as many different types of tools needed, there are as many subcategories, each with a specialty. The tool I thought was properly referred to as a shovel is, in fact, a spade but it could just as easily be confused with a trowel. These tools are of different sizes, but are similarly shaped, the trowel being the smallest and in my mind the most practical for the gardener. A trowel is basically a small spade. It's ideal for weeding and working with small plants like bulbs, vegetables, flowers, and herbs. All three are used for digging holes, but a shovel is used for moving the dirt or anything heavy.

With more than one way to dig, I often wondered if I was using the right tool and in conjunction with that thought, how many tools did I really, truly need? Once again, I consulted my gardening books. Fortunately, several writers have enumerated their preferences in detail. I was happiest when I came upon Anna Pavord's recommendation: "The basic kit is the same one that John Evelyn described in the seventeenth century: spade, fork, trowel, line, hoe, rake, something to cut with. For him that would be a beautifully made pruning knife. For us, it's a pair of secateurs." I've been using secateurs—good, old-fashioned pruning shears—for the past thirty years to cut back the bushes in the garden on Long Island. Perhaps because of my comfort level with them, I'd put secateurs, along with a trowel, at the top of my list. Derek Jarman would vote for a rake, hoe, shovel, trowel, and secateurs, though not necessarily in that order. He considers spades and forks of little use except for shifting manure. Katherine's go-to tool is a short garden spade with a narrow square blade she grew up using in her grandfather's vegetable garden. Her second pick is a garden trowel with a pointed tip, perfect for both planting and weeding. Jamaica Kincaid says that the first hand tools she acquired were a trowel, dibble, cultivator, and weeder. I've come to value them all, particularly the dibble, which is, at its most basic, a pointed wooden stick used for making holes in the ground for seeds.

Weeders and cultivators are useful for forcing weeds—what I think of as plant survivors—out of the ground. Monty Don additionally recommends a hoe, especially for vegetable gardens. Katherine, Gaye, and I, and my bet is Kincaid and Don, too, would rather pull weeds with our hands when they don't resist us or even poke holes for planting seeds by hand, if we have the time. There's something satisfying about pulling a weed out of the ground with its roots still intact. It's equally frustrating to pull a weed out without its roots as chances are it will reappear soon.

Then there's the matter of where to put your tools when you're working. It's useful to have a canvas bag—one you don't mind getting wet and dirty—for carrying around your most-needed tools, along with an assortment of odds and ends, depending on your taste. I also keep the following items in my bag:

- Twine: for tying plants up and marking out rows.
- Scissors: for cutting the twine—and anything else that needs it.
- Pen and paper: jotting down thoughts as they come to me is so much better than trying to remember little details at the end of the day.
- Labels: for noting essential information, including plant names. While useful, many gardeners find the plastic labels that come with the plants unattractive. I tuck them into the dirt at the edge of the bed to hide them from sight. I prefer wooden labels on which I can write the details I need, and put them at the end of the rows, highly visible and feeling to me more like a working garden.

While I'm developing a friendly familiarity with my plants, I still forget some names. I've taken my social cues in the garden from Celia Thaxter and Prince Charles, both of whom famously talk to their plants. When speaking out loud to your plants, it's good to be on a first-name basis with them.

DRESSING FOR YOUR GARDEN

This is Herb, my fair-weather friend. He settles in around Memorial Day and after the last harvest heads south (to our garage) for the winter. You may have noticed he has both fall and spring shirts. Funny how an orange plaid flannel looks so off in spring. But he always wears brown corduroys in honor of his namesake, Herb, a character in Reginald Arkell's *Old Herbaceous: A Novel of the Garden*. The fictional Herb was given a pair of corduroys upon becoming a gardener at an English estate with a stately garden. He didn't like corduroys: "That was the trouble with corduroy—it lasted forever. . . . Once corduroy had you in its grip, it never let go." I, for one, like corduroy, and found Herb's pants inexpensively online. So far they've lived up to their reputation; although a bit faded by the sun, they still look brand-new.

Roy Strong also likes corduroy, saying that, for years, he had "a drawer for garden clothes and for years it's been the same corduroy jeans for winter and cast-off blue jeans for summer." When people visit, he advises that they "come dressed like peasants." He continues, writing, "That is always my injunction to anyone coming to see the garden, for peasants they will seemingly meet as my wife and I stand in our gardening clothes to welcome them. Gardening clothes are special. . . . On the whole what one wears are cast-offs. . . . My pile looks like a male fashion throwaway from the sixties and seventies. For years I wore a wonderful pale cream corduroy jacket with pockets with huge flaps and a broad belt. . . .When that fell to pieces I donned a donkey-coloured Aquascutum overcoat, a relic from the same era."

I follow his advice on boots and shoes, items in his mind that "you can't have enough of." In the summer, he favors discarded sneakers or slip-ons. In the winter, he wears Wellingtons, but warns they don't last, writing, "Sooner or later you peer down and they've split. Worse, they spring a leak while you're wading through mud. That is why you can't have enough of them."

I'm a Muck Boot fan. Growing up, my favorite possessions were my baseball mitt and my Shark water ski. Now it's my trowel and my Muck Boots, the most

comfortable of the various Wellingtons I've tried. However, unlike Hunter boots, they leak, a lot, if you step in a pool of water.

Gloves and hats are the other essentials; trial and error is the only way to know what fits best. While gloves are great when the sun is glaring or it's very cold, I still try to rely on my bare hands whenever I can. A hat should either stay put on your head or have a chin strap to keep it in place. I'm dedicated to hats.

E. B. White took an interest in gardening, especially as it was a beloved pursuit of his wife, *The New Yorker* editor Katharine White. He compared her clothing to Gertrude Jekyll's, whose boots were immortalized by Sir William Nicholson in his painting *Miss Jekyll's Gardening Boots* (1920), shown above.

White wrote:

When Miss Gertrude Jekyll, the famous English woman who opened up a whole new vista of gardening for Victorian England, prepared herself to work in her gardens, she pulled on a pair of Army boots and tied on an apron fitted with great pockets for her tools. Unlike Miss Jekyll, my wife had no garden clothes and never dressed for gardening. When she paid a call on her perennial borders or her cutting bed or her rose garden, she was not dressed for the part—she was simply a spur-of-the-moment escapee from the house. . . . Her Army boots were likely to be Ferragamo shoes, and she wore no apron. I seldom saw her *prepare* for gardening, she merely wandered out into the cold and the wet, into the sun and the warmth, wearing whatever she had put on that morning. Once she was drawn into the fray, once involved in transplanting or weeding or thinning or pulling deadheads, she forgot all else; her clothes had to take things as they came. . . .

She simply refused to dress *down* to a garden: she moved in elegantly and walked among her flowers as she walked among her friends—nicely dressed, perfectly poised. If when she arrived back indoors the Ferragamos were encased in muck, she kicked them off. If the tweed suit was a mess, she sent it to the cleaner's.

Like many gardeners, I hold the garden in reverence, and therefore care about what I wear. One day, when I was meeting Katherine and Gaye, I purposely wore a very old pair of white jeans, because they had lost their shape and were falling down, and I doubted I'd wear them again. You should have seen their faces. They were clearly trying to be polite and not say anything: Who wears white in the garden? We laughed about it later. I tend to wear old but favorite clothes: jeans I love, cozy sweaters that don't itch, and my favorite jackets. Trying to match the right jacket to the weather will be, I have a hunch, a lifelong pursuit. But really, as to what to wear in the garden, it's your garden. Obviously, anything goes.

V. SEASONALITY AND SUSTAINABILITY

When you look outside at a downpour, do you think, "Oh hooray—good for the garden"? . . . Do you, when you notice that the downpour has passed quickly, feel a sense of disappointment? Then you are a gardener: the highest point to which humankind has evolved.

CHARLOTTE MENDELSON

LUCK AND SUSTAINABILITY

Beverley Nichols believed that there are four must-have qualities for any good garden, writing, "Light in a garden is a quarter of the battle. Another quarter is the soil of the garden. A third quarter is the skill and care of the gardener. The fourth quarter is luck. Indeed, one might say that these were the four L's of gardening, in the following order of importance: Loam, Light, Love and Luck."

In fall 2019, after the fence had gone up and I'd done my first planting, I experienced luck in the garden on a day that's historically considered unlucky: Friday the thirteenth. Not only that, but there was also a full moon. A full moon coinciding with Friday the thirteenth is a rare occurrence; the two won't be in sync again for another thirty years.

Strange things happened all day to the members of my family, from minor mishaps to a potentially dangerous accident, so a sense of worry lurked in the back of my mind. Fortunately, all was well by evening; when the children, Don, and I were together in Connecticut, safe and sound, I was relieved.

Once we were home, I went to check on the rows of lettuce we'd planted just a month earlier. To my surprise, they were up! What luck! It was as if while we were gone for only a few days, the lettuce had suddenly woken up, their leaves open to the sky. The peas stood to attention and the radishes poked their red bodies out of the ground. The cabbages were as big as any I'd imagine Peter Rabbit coming across. It was the harvest moon, a time when plants are at their fullest, and ours looked it. Everything was ready to be eaten. Our dinner that evening became a feast—our first harvest of our first season of plants. The produce was so fresh, it was best eaten with just a quick blanching and a touch of oil and seasoning. I often reflect on this happy memory, a warm fall day when, for me, the house felt a bit more like a home, our family together in the kitchen and the garden full of life.

If you think about it, the gardeners are first in our food chain as without the gardener, the cook wouldn't have ingredients. Many gardeners become cooks, and vice versa, so they can more directly enjoy the fruits of their labors. They can also be at cross-purposes, however. Amanda Hesser, who spent a year as a cook in a Burgundian château, thinks that the "relationship between cook and gardener has traditionally been a sour one. The cook always wants vegetables a month before they are ready, whereas the gardener would much rather deliver the world's largest carrot to the kitchen." While that may lead to wars in the kitchen, both are operating under the same principle: seasons rule. Unfortunately, today we've found all sorts of ways to cheat the seasons.

When I was growing up, we were excited when asparagus season arrived and we could enjoy its fresh spears or vividly green peas almost every night until it was time to move on to broccoli and cauliflower. And then the tomatoes or locally grown corn in summer, and pumpkins in late October. Too many restaurants today have brussels sprouts on the menu all year long, a vegetable I think of as being for a meal like Thanksgiving in late fall. You can get squash or pumpkin soup in May and peonies and tulips in December. If we as consumers and home cooks thought a bit more like the gardener, perhaps seasonality would come back into fashion. I've certainly come to understand this idea in a much more immediate way having now grown my own vegetable garden.

Alain Passard, renowned chef and owner of L'Arpège, a three-Michelin-star restaurant in Paris, knows seasonal produce is not only the most beautiful, but also the best tasting, saying, "Nature created the most beautiful recipe book in the world—and it changes every three months." He believes this so deeply that in 2001, he made a radical change, deciding to showcase vegetables on his menu instead of offering meat. It was a risky move that made for a few difficult years, but his customers eventually came to appreciate his plant-based cuisine. Today, he grows his own organic vegetables in gardens outside Paris for all the restaurant's meals. The produce is harvested the same day it's served so the customer gets to enjoy meals at their absolute freshest; it's as close to eating straight from the garden as it can be. The terroir of each garden is different, so the produce is, too. The same vegetable can have different flavors depending on where it was grown.

The importance of seasonality reminds me of M. F. K. Fisher, who wrote about the simple joy of eating straight from the vine. She said that the "best way to eat fresh [peas] is to be alive on the right day, with the men picking and the women shelling, and everybody capering in the sweet early summer weather, and the big pot of water boiling, and the table set with little cool roasted chickens and pitchers of white wine. So . . . how often does this happen?"

There's been a global revolution in food since Passard's menu change, with gardeners and cooks embracing a more sustainable approach to dining. It's happened in America, with Alice Waters leading the way; in Italy, where the Slow Food movement was first created; in the notably agriculturally proud France; and in England, where, for example, the huge kitchen garden at Gravetye Manor produces all the vegetables for what was once William Robinson's former home but is now a country-house hotel and a first-class restaurant.

The kitchen simply must work with whatever the garden yields. Certainly one of the deepest pleasures of gardening for me has been the joy of working in harmony with nature and the seasons. Passard says his "pet hate" is "chefs who do not respect seasons." Eating a tomato in December is perhaps the biggest sin. He says, "I hope that we will return to respecting nature and putting seasons back where they belong. . . . That would be the most futuristic idea."

A working garden is a gift to anyone who likes to cook and make the most simple ingredients sing. Antonia Read taught me how to make coleslaw. It's a straightforward recipe; the cabbage is the star. Despite now having the recipe my dish can't compare with hers.

ANTONIA'S COLESLAW
Serves 6

A good coleslaw begins with a ripe head—or two or three—of cabbage, and in this department, I'm happy to say that our garden has yielded a fresh starter ingredient. In fact, I planted cabbage just for the pleasure of having it used in her recipe. After our first harvest in 2019, Antonia included this dish as part of our family feast.

INGREDIENTS

For the salad

- 2 pounds green cabbage
- 1 medium carrot, grated
- 5 sprigs of parsley with stems, finely chopped

For the dressing

- 3 generous soup spoons extra-virgin olive oil
- 1 teaspoon white sugar
- ½ teaspoon black pepper

½ teaspoon salt

2 soup spoons white vinegar

1 soup spoon mayonnaise (Do not use organic mayonnaise.)

PREPARATION

Take the inner white leaves of the cabbage only, plus two green outer leaves to add a little color to the dish, and chop finely with a knife. (Don't use a mandoline, as it doesn't give the coleslaw the same texture.) Place in a medium-sized bowl and set aside.

Place the grated carrot and chopped parsley together in a small bowl. Add the ingredients for the dressing, then whisk everything together until well blended. For a crisp coleslaw, add the dressing to the cabbage and refrigerate, covered, for two hours before serving. The coleslaw is a little softer and often tastes even better the following day, but it is best eaten within a day or two of preparing it.

A FULL CYCLE

Three months after my Lucky Day—my first harvest—in 2019, came the saddest day, Don's death in December, something I still find hard to believe. I didn't get much gardening or planning done in those two endless, cold months, returning to the garden in February. My sense of possible renewal was short lived when, as with many other families and communities in the New York region, my family and I went into quarantine in mid-March. Fortunately, we'd already made it through February, the month Joseph Wood Krutch calls the "3 a.m. of the calendar." He writes, "Spring is already too far away to comfort even by anticipation, and winter long ago lost the charm of novelty." But March, of course, is still winter, and it felt like it. Gray skies predominated. Frosts came almost nightly and lasted until May, a phenomenon rarely seen in Connecticut. In fact, despite its New England backdrop, southern Connecticut is a rather temperate place.

I was deeply fortunate to be in such a natural, lovely place for the long, fearful months of lockdown. Surrounded by nature, I could use some of the skills I'd been developing—observation, planting techniques, patience—every day. During a time of so much global uncertainty, I could get my hands in the dirt—something tangible that helped to anchor me during a time that felt decidedly unreal. Literally face-to-face with nature, I watched spring unfurl by the hour. The garden became my calendar, the way I marked the passing hours and days. Days merged into weeks, and time seemed to move more quickly than normal. The garden gave me an equilibrium I'm sure I wouldn't have otherwise had. I often wondered if I had been a more experienced gardener, would I have been able to tell whether it was the end of March or early April without looking at my watch or a calendar?

I focused on the garden, working through layout options and plant choices until I felt I'd achieved a good balance between vegetables and flowers. Every plant has its purpose (vegetables, beneficials, and pollinator flowers), and they all work together as a whole. In making my choices, I came to the hard realization that I couldn't have all I wanted. So, I scrapped some ideas—planting a special pollinator shrub bed, for instance—and adapted others, deciding to create a special raised bed for the dahlias outside the fence rather than keeping them where they'd been. Perhaps more importantly, though, I followed Roy Strong's example and brought some color into our lives. "Even on the dreariest winter's day," he writes, "my spirits are lifted by this joyful splodge of colour welcoming me. . . . As you look through your window there must be something that catches your eye ready for a quick lick of colour. You'll never regret it."

With his words in mind, we painted the cold frames a happy yellow. His advice is spot-on. Even though the frames are a small spot of color, they do look cheery against our gray shingles, and are so much better than plain old wood. They also remind me of Derek Jarman's colorful window and door frames, which I

OPPOSITE: *Hilma af Klint,* Primordial Chaos No. 24, *1906–1907. The happy, vibrant yellow rose in this painting echoes Roy Strong's belief that "even on the dreariest winter's day, [one's] spirits are lifted by [a] joyful splodge of colour. . . ."*

imagine people can see in the distance even through the English gray mist, and what a happy sight that would be if his cottage is your destination. I made another colorful upgrade, painting the door frames of our garage bay converted into a work/potting place a bright grass green, the same hue Monet painted his own door and window frames. Each time I enter it, I admire my "Monet green." These springtime colors were a salve on our otherwise cold and dreary winter in Connecticut.

I've often felt that seasons and color are intertwined. Without being religious, I wonder if some divine hand is at play, coloring the leaves and flower petals in their seasonal palette. Diana Athill writes, "The spring is largely yellow, a golden time, followed by blues and purples of foxgloves and forget-me-nots." By fall, these colors, having exhausted themselves in summer, loosen up, with Lauren Springer writing, "What may jar in the May and June garden is a welcome sight in October. Colors have richened and deepened with the cooler temperatures and golden light." Winter, of course, brings white, a color Jamaica Kincaid says only "makes you feel the absence of color."

Kincaid also observed several years ago in her Vermont garden that "March came to an end in mid-May." That's exactly what happened in Connecticut. While spring officially begins on March 21, for many, springtime has a different start date, one that's fixed in spirit. For me, it's when my crabapple tree blooms.

At the end of March, we looked around and saw only brown and gray branches and skies. This drudgery continued throughout April and well into May until I thought spring was on lockdown, too. The flowers of the cherry trees taunted us in mid-April, making us think that their leaves would be soon behind. It took another month for them and other leaves to appear, and even then we could have blinked and missed the whole show. Our trees went from having nothing to bearing a full-blown green canopy in only a few days.

As the world shifted and churned around us, so did the garden. Each day brought noticeable changes and surprises. For me, the most apparent and wonderful were the tulips. The contrast between their vivid colors and the dreary, dark skies, both literally and figuratively, was stark. We all appreciated them, my children perhaps even as much as I. I now understand the rage of tulipomania: their beauty is spellbinding.

Springtime is typically a hopeful season—a time when the promise of a better day feels like it's just around the corner. Gardeners are naturally forward-thinkers, a sentiment that helped me get through the rocky winter and the early gray days of spring. Penelope Lively says that she's still gardening for a future at the age of eighty-three, writing, "the *Hydrangea paniculata* 'Limelight' I have just put in will outlast me, in all probability, but I am requiring it to perform while I can still enjoy it." For me, this is evidence of our endless capacity for hope—something I wanted to believe in. The tiny plants of our spring Big Plant grew and grew until the garden was filled with an abundance of color and texture, each flower, vegetable, and herb celebrating the return of the summer sun. This Big Plant is a day when we bring in lots of plants and some seeds, and lay them out, moving pots here and there. We clean up everything that's had its day and plant everything else. It's a rewarding

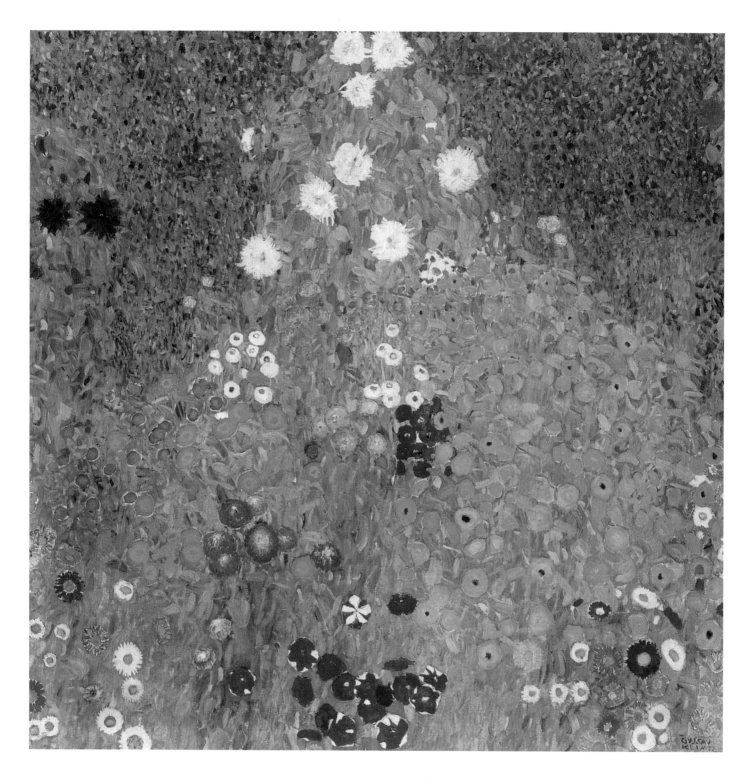

ABOVE: *Gustav Klimt,* Flower Garden, *1907. It's remarkable how much this painting and our midsummer garden (*OPPOSITE AND PAGES 192–193*) mirror each other. The similarity in color and composition is uncanny.*

day's work, especially fun to do with Katherine and Gaye, and it gives me time to think about what I'd like to do in the garden next season. We did two Big Plants during the year: in the spring and fall.

With summer's arrival, we said goodbye to our fresh "spring greens"— the lettuces, spinach, and peas, which I had found almost miraculous. My second harvest—in the early summer of 2020—was a big success, with an abundance of produce: kale, beets, zucchini, turnips, carrots, spring onions (they were so delicious, I added them to most dishes), basil, cabbage, currants, lemon verbena, and lemon balm. We had such an overabundance that Gaye was able to take several bags' worth of produce to her local food bank.

Then my garden had what I'm calling the midsummer blues. After a long dry spell, a huge storm with gusty winds arrived that nearly flattened every growing thing—not just in our garden but in our neighbors' gardens, too. Next, the plants began to wilt from the heat and then, due to our newly installed but faulty irrigation system, were overwatered. Even with these setbacks, my little garden still produced beautiful artichokes, beans, tomatoes, arugula, blackberries, grapes, and even more kale, beets, and turnips.

By September, the sun began to set noticeably earlier and lower and it was becoming hard to eat dinner outside. One fall evening, I realized that my sense of color had shifted: The hues of everything around me had seemed to deepen. Karel Čapek calls this month the time of "ripening wine," and I can understand why. In the fall, nature's mood changes. The light becomes more amber. Suddenly, it's the richness of the dahlias and ruby red apples that seize the day.

As I write this, the garden will soon be put to bed for its winter sleep. Today, September 12, 2020, a full year since my first harvest, we had our fall Big Plant. (We seeded cover crops in August and they've already grown so much.) We planted several brassicas: cabbage, cauliflower, broccoli, and brussels

OPPOSITE: *Some vegetables from our summer harvest in the good company of our family dog and a favorite Wolfgang Tillmans photograph.*

sprouts. All enjoy the cool weather so we'll be able to extend our harvest and still enjoy them this fall. We also planted pollinators—brown-eyed Susans, dwarf Joe Pye weed, goldenrod, and echinacea that will bring the bees and the flowers through October. Eighteen months after I first started, the structure of the garden is set, the fence is holding up nicely, the grass paths are amazingly healthy—a nice surprise as I thought for sure they wouldn't last and we'd be debating gravel or brick paths right now—and many plants are thriving. This fall, the cover crops came in well, especially the light yellow-green buckwheat, and gave a springlike freshness to the garden that reminded me of the lettuces.

GARDENS ARE GOOD FOR YOU

In the 1930s, a popular parlor game among the intelligentsia went like this: Of these three aspects of life—climate, friends, or work—which influences you the most regarding where you choose to live? I've asked this question to people over time and the answers largely depend on a person's circumstance. Some say climate right away; but those who don't generally give their answer, then stop and think about it, and return to the importance of climate, often wanting to change their original response. Curiously, I've noticed that most garden writers I've read are from places with very distinct seasons. W. S. Merwin was a prominent exception as his garden was in Hawaii. I've also noticed that most prominent American gardeners are from New England and, of those, the majority are from Connecticut: Eleanor Perényi moved to Connecticut from Hungary, Clare Leighton from England, and Michael Pollan created the garden about which he wrote in the northwestern part of the state.

Climate clearly impacts us all. Not to minimize the serious worries many of us have about our planet's well-being—and naturally, gardeners take climate issues seriously—but it is heartening to think of the benefit one can give and get by simply working a small bit of land. It may not feel like doing so makes a global impact, but cumulatively with other gardeners, it does. One hundred and fifty years ago, Charles Dudley Warner intuited that the "man who has planted a garden feels that he has done something for the good of the world. He belongs to the producers. It is a pleasure to eat of the fruit of one's toil, if it be nothing more than a head of lettuce or an ear of corn." Simply planting a tree is beneficial: A single tree can absorb up to ten pounds of air pollution a year and release 260 pounds of oxygen. To put it in more digestible terms: One tree produces a day's supply of oxygen for four people. A tree can also be home to hundreds of species of insects, fungi, moss, and animals.

Oliver Sacks, the distinguished neurologist and author of popular anthologies of his case studies, often wrote about the deep value of gardens to our health. He reported that, of his patients with chronic neurological disease, there were only two types of non-pharmaceutical therapy that helped: listening to music and being in gardens. Time and time again, he witnessed the restorative and healing power of music and nature to soothe his patients, the results of which were often more helpful than medication.

Nature exerts a powerful feeling deep within us. Biophilia, a scientific-sounding, almost off-putting word, is, in fact, about something profoundly moving: our love of nature and all living things, not just other humans. We have a natural tendency and desire to want to commune with nature. As Henry David Thoreau said, "Shall I not have intelligence with the earth? Am I not partly leaves and vegetable mould myself?"

ON SIMPLICITY AND PLANS FOR NEXT YEAR

I've often felt that I've spent years trying to reach the unattainable goal of living a simplified life . . . sometimes complicating my life in doing so. In so many ways, wisdom comes with age, and it's the older gardeners and writers like Christopher Lloyd, Russell Page, and Rosemary Verey who recommend simplifying both your life and your garden. Page wanted his garden to be manageable enough to be one man's work.

I like his approach and thought it would be attainable when I first started out. I quickly discovered, however, that even a plot of land measuring forty-eight by fifty-four feet requires round-the-clock vigilance and a rotation of able hands. While my vegetable garden, with its linear rows and defined structure, is relatively uncomplicated, the back square beds are more elaborate in their planting. I think I'll simplify the design next year. That said, it's been wonderful to see the unexpected mix of flowers and vegetables that have popped up. To me, their unusual combinations created an almost Klimt-like look in mid-July that wouldn't have happened had the rows and plants been too strictly defined. Of course, it's all a matter of personal taste and dependent on how much work you want to dedicate to creating a simplified life.

As a novice, you want gardening to be fun. You also need to learn about plants, which you can do only by closely observing at least one cycle of blooming. Planting a garden takes time, more than you'd first guess. Monty Don recommends planting only what you love: That's my goal. But if your garden is large—or even slightly bigger than what you can easily maintain on your own—that's close to impossible to do until you get through your first, second, or even third round of plants. Several writers like Vita Sackville-West and Roy Strong encourage you to think big and go bold. And for experienced gardeners, I'm sure that's true. But for a beginner like me, I'd recommend something different. Start small.

I've been thinking a lot about next year's plan and how to balance dreams with practicality. Of course, I must start planting bulbs and buying seed this fall,

so there isn't much time to waste. I'll still have abundant tulips in all the beds, but once they stop flowering, I'll probably put a cover crop over the square beds, and work only on the rectangular ones, which will be much easier to manage myself as they are straightforward rows of plants. I'm eager to try the tulip–dahlia rotation system, most likely in the middle two beds, and will reserve the outer ones for vegetables. The lettuces did so well in the more shaded corner bed that I'll copy that.

Self-sufficiency—the ability to sustain my family's food sources—if life ever came to that and I highly doubt that it will, is nonetheless becoming ingrained in my system. Certainly not as deeply as it was in the Founding Fathers' days, but it's a force that is growing within me. More than a practical concern, I think it's partly because of my stronger connection to the land, and my desire to work with it to make something productive. This also ties into the concept of today's victory gardens, and the sense of community among the more and more people who've been finding satisfaction in their newfound interest in sustainability through the food they grow.

So all the vegetables we love—from my lettuces to Serena's kale, and William's willingness to try anything new—will be there. While this year's peas were a flop, pretty to look at but not good to eat, I'm going to try again. The ability to look forward, to be excited about what's to come, is an essential character trait of the gardener, and amazingly it's becoming one of mine. In gardening, there's always next year. If you plant something and you don't like it, you can fix it next time around. Helen Dillon's motto, "Buy it and see," is now my motto, too.

As I was typing, with only crickets and a loud frog outside to entertain me, I had a nagging urge to go check the garden. I'd been there only a couple of hours earlier, and had a lot of work still before me, yet its call was strong, so I went and wandered. I looked around and savored its lushness—such a bounty of beauty and nourishment—that was there in late dusk. As Perényi promised, the scents were stronger than they'd been just a few hours earlier. I've continued to learn, and realize I will be doing so, for as long as I have a garden. Where I started off learning to garden, I now garden to learn.

VI. A NEWFOUND WAY OF BEING

People often ask me what is the most important thing that I've done with my life and I reply, 'To have made a garden.' . . .if you also asked what entrances me most about gardening. . .the answer could be given in one word: hope. Every winter we live in hope that spring will come. This or that doesn't do as well one year as last but the hope is next year that it will do better. We plant in hope. We live in hope. What more can anyone ask of life?

SIR ROY STRONG

Throughout this project, I was impatient to decide if I'd become a gardener. I wanted to know where I'd land, and whenever the impulse to judge occurred, I'd call on an African proverb I had wedged into my mind, "Don't judge the day until the night."

After eighteen months, from building the garden to seeing it through several seasons, I still consider myself an urban dweller. Yet, I'm certainly more closely tied to the rhythms of nature than I've ever been. And I have no doubt that were I to be fully immersed in urban life, I'd deeply miss my newfound friendship with the natural world.

Do I feel more rooted? This is an unfair year to judge, as 2020, with the coronavirus and all the hardship accompanying it, has been uprooting for us all. But, even with the upheavals, I'd say yes. Seeing those brilliant tulips every gray day, which seemed to holler, "Spring is coming; it really is!" was like a bolt of sunshine, but stronger. The hopeful feeling of spring fortified me through the long nights. It stays with me now even when fall is calling.

I can also say with certainty that the garden has increased my sense of well-being and happiness. I think this comes in part from the fact that gardening, being such a tactile activity, taps into what positive psychologists call "flow": a state of complete immersion in an activity. In this mental state, time seems to fly; I've experienced this firsthand. Every action, movement, and thought seamlessly follows from the previous one. When I'm gardening, all my skills are engaged and my whole being is involved—something the Swedes call "work joy," a concept that instinctually makes sense. Charlotte Mendelson also flagged the concept of "flow." She called it "the secret of life, or at least of contentment . . . enabling us to live purely—and contentedly—in the moment." I couldn't agree with her more.

Over the better part of my life, I'd read about the ways gardening changes your existence. And for at least two-thirds of my eighteen-month-long project, I doubted it would change mine; the benefits seemed too good to be true. At a certain point during the spring, however, while taking in the colors of the tulips, I happily noticed a growing change: I felt more at ease in the garden and more sure of myself there. Now I wonder if I'd been too self-critical. At the end of the day, I am someone who gardens, and that's enough for me. A garden is a living embodiment of your involvement with life.

Being a gardener isn't about having a greater knowledge of plants or savoring the choice of lettuce at the nursery. It's your imagination, memories, dreams, and life experiences made tangible. As Beverly Nichols writes, "you can no more stop a garden from walking, in spirit, into the house of a gardener than you can stop the sea from flowing, in spirit, into the house of a sailor. . . .

For there is always a little mud on the floor, a feeling of flowers everywhere, a perpetual surge and whisper of branches, and a heavenly scent of flowers." It's the taste of fresh oranges and tomatoes on the vine, the scent of lilac in spring, and apples in early fall. It's the steady reminder that life can be good even in the midst of heartbreak and sorrow. The future is uncertain. It's best to grow what you can when you can.

Would I advise you, my fellow intrepid gardener, to begin a project of your own? My wholehearted answer is "Yes." It will bring you life.

APPENDIX

ANNUAL TO-DO LIST

Henry Mitchell wrote, "Gardening is a long road, with many detours and way stations, and here we all are at one point or another. It's not a question of superior or inferior taste, merely a question of which detour we are on at the moment. Getting there (as they say) is not important; the wandering about in the wilderness or in the olive groves or the bayous is the whole point." I've taken more than my share of detours on the path toward becoming a gardener over the past eighteen months. Thankfully, most have been enjoyable. So I don't get too far off, however, I keep monthly to-do lists and, once a season is done, assess my successes and mistakes, hoping that my observations will help me the next time around. Hopefully, these monthly lists will help you organize your own yearly plans.

January

- Order cold frames.

- Complete building gates and install them.

- Narrow down inspiration for planting and design ideas to be more realistic.

- Review initial plant list of vegetables and flowers.

- Order flower seeds.

- Start to organize planting list and allocate flowers and vegetables to beds.

February

- Explore various ideas for planting layout of beds.

- Order dahlias.

- Order more seeds for flowers and vegetables.

- Source and select growers for growing seeds.

- Order seed-growing supplies.

- Explore pollinator bed idea and its location.

- Create design plan.

- Consider all options to repair or improve looks of blackened fence rails.

March

- Pollinator planting design/plants sent to contractor for sourcing.

- Finalize planting plan.

- Paint fence a dark green.

- Decide pollinator beds are too much to take on this year.

- Source crabapple tree from Rosedale Nursery, my source for many plants.

- Plant pansies.

- Paint cold frames yellow; plant with lettuce and herb starts as still too cold to plant in ground.

- Begin loosening the soil in areas where vegetables will be planted.

- Order any heirloom plants not available at nurseries, including sweet peas, scented geraniums, most medicinal herbs, online.

April

- Revise planting layout in square beds.

- Source flowers at nursery and make any necessary substitutions.

- Turn on irrigation.

- Feed lawn.

- Lay out new dahlia bed plants on paper.

- Straighten grass edges.

- Bring seedlings outside to harden off.

- Check with growers on seed starts.

- Test bed soil in lab.

- Review bed design and adjust planting and orders.

- Move starts from cold frames to garden and replant cold frames.

- Plant early cold-hardy vegetables like spinach and peas (very cold spring, so plant on the late side).

May

- Make notes for next year's tulips.

- Plant sweet pea starts (late this year because so cold).

- Pull tulips and turn over soil with broad fork to add air into soil.

- Check in with growers/nurseries to make adjustments for plants that could not be found or failed to grow.

- Spring Big Plant! Plant virtually the entire garden as only a few cold-hardy vegetables had previously been planted between the tulips.

  - PLANTINGS IN THE RECTANGULAR BEDS: chives, onions, leeks, beets, tarragon, cauliflower, red and green cabbage, mixed varieties of lettuce, broccoli, kale, oregano, basil, and thyme. Flowers include cosmos, dianthus, bachelor buttons, pink geraniums, and five kinds of poppies.

  - PLANTINGS IN THE SQUARE BEDS: leeks, dill, lemon verbena, artichokes, bronze fennel, basil, and angelica. Flowers include hollyhock, yarrow, delphinium, dianthus, sweet William, strawberries, scented geraniums, nasturtium, and butterfly weed. In the back orchard bed, planted tomatoes, red currant, sweet peas, and lady's mantle.

June

- Cut back late-blooming Cheerfulness daffodils (inside back bed).

- Plant evening primrose in back bed after daffodils are cut back.

- Harvest: lettuce, radish, turnips, beets, broccoli, last of spinach, and kale.

- Cut back chives and dianthus.

- Train morning glories, roses, blackberries, sweet peas, and tomatoes along the fence.

- Harvest the last of the peas and pull them.

July

- Harvest: turnips, beets, onions, kale, cabbage, strawberries, currants, and the last of the lettuce.

- Thin and trim the morning glories, which are growing like weeds.

- Pull out the poppies after the blooms finish, some morning glories, yarrow.

- Pull and reseed bachelor buttons.

- Install dahlia stakes and weave the support.

- Plant early fall crops as small plants (not seeds): pumpkins, acorn squash, delicata squash, and butternut squash, with the goal to harvest at end of October.

- Continue to train morning glories and climbers on fence.

- Plant more cabbage, kale, and cauliflower for fall.

- Thin out lemon verbena, lemon balm, scented geraniums, coreopsis, valerian, fennel, and nasturtium.

- Deadhead: butterfly weed, toadflax, and coreopsis.

- Continuously harvest, deadhead, and stake sunflowers.

August

- Pull some apples so others grow bigger.

- Remove basil infected with a summer fungus.

- Fertilize dahlias.

- Start spraying apples with deer repellent as deer have discovered them.

- Cut back lavender and hang lavender to dry.

- Harvest artichokes; leave some to flower to add color to the garden.

- Plan fall pollinator and vegetable planting.

- Research best cover crops for improving drainage and seed. Begin to plant.

- Continue to train dahlias as they grow.

September

- Plant cover crops—buckwheat, crimson clover, oat, and pea mix—in the bare spots.

- Fall Big Plant: Plant fall crops and pollinators. Crops: cabbage, cauliflower, broccoli, and brussels sprouts. Pollinators: brown-eyed Susans, dwarf Joe Pye weed, goldenrod, and echinacea.

- Thin squash and pumpkin patch leaves. Remove dead leaves to make some space for the sun to reach vegetables.

- Harvest lettuce and kale.

- Pick blackberries and tomatoes.

October

- Put down straw over vegetable beds (to keep fall crops dry in rain).

- Pick apples.

- Order tulips.

- Turn off irrigation.

- Cut back perennials.

- Move fig trees to garage to protect over winter.

- Gather flower seeds for next year's seedlings.

November

- Harvest last of the frost-resistant vegetables: turnips, cabbage, squash, and kale.

- Leave cold-hardy herbs—parsley, rosemary, and lemon balm—to come again in the spring.

- Pull up all vegetables that succumbed to frost and use as compost.

- Cut back dahlias and labels and dig up tubers to store in cool basement.

- Turn and aerate soil in garden.

- Plant tulips.

- Prune grapes and blackberries and mound the base with straw or mulch to protect through winter.

December

- Add an additional layer of straw to the beds for added insulation.

- Clean and sharpen tools.

- Prepare for the year ahead.

Gustave Caillebotte, The Gardeners, *1875–1877.*

Duncan Grant, The Doorway, *1929. Grant, a member of the Bloomsbury Group, was a very good friend of Vita Sackville-West. Sackville-West, the inspiration for the character Orlando in Virginia Woolf's novel of the same name and a world-renowned gardener, created Sissinghurst, one of England's most famous gardens.*

LITERARY MENTORS IN THE GARDEN

Long before I ever set foot in my own garden, I sought out gardening advice, opinion, and insight from authors both known and unknown. Over time, I've come to think of these writers as my literary mentors. While I gleaned useful information from every author cited in this book, some made an especially indelible impression. As I read their writings and worked on my own, I learned more about the lives of these fascinating writers and gardeners, which I share with you along with some reading recommendations.

ELIZABETH VON ARNIM (1866–1941): A British novelist who spent many years in Germany, she wrote about her garden largely through her transparent semi-autobiographical novels. Her garden clearly meant a great deal to her, as she signed more than twenty of her novels as "by the author of *Elizabeth and Her German Garden.*" She chafed under the social mores of her station in life, where it wasn't appropriate for her to do the physical work in the garden herself. I sense she would have loved to be a dirt gardener.

KAREL ČAPEK (1890–1938): A prolific Czech novelist, playwright, and essayist, Čapek's classic *The Gardener's Year* (1929) is very accessible, in good part because of its humor. The book is beautifully written and has been cited as being as close to a literary masterpiece as a book on gardening can get. It's an appealing read for inexperienced and experienced gardeners—and nongardeners, too. While his writing style is lighthearted, he weaves in deeply thought-out principles by which one should think of and understand gardens. Čapek is revered by many for his knowledge of soil, and his belief that the worth of any garden begins with good soil. Like many garden books, *The Gardener's Year* is narrated month by month. His book includes charming illustrations by his brother, Josef. To this day, Čapek is still a hero to many, specifically because of his focus on the backbone of the garden—the soil.

MARGERY FISH (1892–1969): While she'd been a journalist, Margery Fish became well-known after the publication of her book *We Made a Garden*, her account of the garden she created with her husband in Somerset, England. The book is both a portrait of building their country cottage garden together and a commentary on their different approaches to gardening. It launched a new career for her; she went on to write several more books on gardens. She became a highly influential proponent of the cottage garden style as well as a serious plantsperson with a special love for hellebores. *We Made a Garden* is one of the

Modern Library paperback reprints of eight gardening book classics, selected and edited by Michael Pollan. I've read almost all of the books in the series and highly recommend them. The last lines of *We Made a Garden* moved me greatly: "I could go on and on. But that is just what gardening is, going on and on." Her husband, when asked when he expected to finish his garden replied, "Never I hope." And that statement, I think, applies to all true gardeners.

ROBIN LANE FOX (1946–): While Lane Fox is an English classicist and historian, and an Oxford don who derives great pleasure from running the garden at New College there, he is most readily recognized for his weekly gardening column in the *Financial Times*, which he's now written for fifty years. I first began reading his column more than twenty years ago, and have been learning from him ever since. While informative, his writing is also humane, straightforward, often witty—never dull—and always reflective of his great love for gardening and respect for nature. *Thoughtful Gardening* is a compilation of eighty of his essays.

ROBERT POGUE HARRISON (1954–): Although I first read Harrison's profound *Gardens: An Essay on the Human Condition* almost a decade ago, its deep meaning has become more essential to me as I have worked on my own garden. In this book, Harrison, a professor of French and Italian literature at Stanford University, links both real and imaginary gardens across centuries to demonstrate humans' need to cultivate and care for gardens. The book also explores the restorative, nourishing role gardens play in our lives. Cultivation of the garden begins with its soil, and the pleasure and value Harrison finds in Karel Čapek's devotion to nurturing his garden's soil is a joy to read.

GERTRUDE JEKYLL (1843–1932): One of England's most legendary writers and garden designers, Jekyll was a pioneer in the influential twentieth-century cottage garden style. She created more than four hundred gardens in the United Kingdom, North America, and Europe. She wrote more than one thousand articles in addition to her numerous books. Her writing was intimidating to me in the beginning; it still is. But once you have some gardening experience under your belt, her books are essential reading. Despite developing increasingly poor eyesight as she aged, she managed to fill her books to the brim with technical advice and moving observations on gardening.

JAMAICA KINCAID (1949–): A superb example of a highly distinguished writer who also gardens, Kincaid discovered gardening by planting seeds in the middle of her front lawn in Vermont. Her works include fiction, non-fiction, and many essays written over more than twenty years for *The New Yorker*. *In My Garden (Book):*, she describes her education in gardening, as well as her own personal history on Antigua where she grew up. Through it, she taught me both essential aspects of gardening as well as a deeper appreciation for the work and rewards involved.

CLARE LEIGHTON (1898–1989): Leighton, an English-American artist, writer, and illustrator, was encouraged to develop her artistic talents by her family. While attending college in England, she studied both painting and wood engraving, for which she is best known. Around age forty, she moved to the United States, became an American citizen, and ultimately settled in Connecticut. She wrote with great praise of the virtues of the countryside and of the people who worked with the land. Her books, such as *Four Hedges: A Gardener's Chronicle*, about the creation and development of a garden from an English meadow, combine both text and beautiful engravings. Her woodcuts bring to life the strength of the workers they portray.

PENELOPE LIVELY (1933–): I came to know the work of this British writer of books for adults and children through her memoirs and Booker Prize–winning novel, *Moon Tiger*. Her focus on both memory and imagination translates into great insights in her sole book on gardening, *Life in the Garden*. Lively's thoughts on gardening are wide ranging and entertaining; they have helped me understand the meaning of the garden in one's life.

CHARLOTTE MENDELSON (1972–): Like virtually all the writers herein, British novelist and editor Charlotte Mendelson writes about many subjects beyond the garden. Yet in *Rhapsody in Green: A Novelist, an Obsession, a Laughably Small Excuse for a Vegetable Garden*, her one book on her tiny garden, she demonstrates a healthy obsession with gardens, taking the reader through a full year. This book is encouraging to all who have little space in which to garden.

HENRY MITCHELL (1924–1993): The author of "Earthman," a long-running gardening column in the *Washington Post*, Mitchell was one of America's most beloved garden writers, in part because he was so adept at making horticultural information so entertaining. His interest in gardening began when he was a boy and old enough to make his own garden, as he was "passionately fond of flowers." I found myself returning again and again to his gardening book, *The Essential Earthman: Henry Mitchell on Gardening*, which is a collections of his columns, whether to mull over any number of his philosophical insights or to seek out his advice on how to best weather the garden in the winter.

BEVERLEY NICHOLS (1898–1983): A prolific author, playwright, and journalist, Nichols wrote more than sixty books and plays, including twelve on his own gardens. I first became entranced by his books twenty years ago or so. While some may find them dated, I think they are charming and humorous. Although his writing encompasses both nonfiction and fiction, including many mysteries, he's best known today for his gardening books, which portray the joy as well as the trials and tribulations of gardening, mostly told through funny anecdotes. His books, especially *Down the Garden Path*, took me away to a gentler era that doesn't exist anymore.

RUSSELL PAGE (1906–1985): Page was one of the twentieth century's most acclaimed landscape designers, in part because he was known for his many gardens in America as well as England. While the gardens he designed were generally expansive and grand, his humble aspirations for his own garden moved me. Largely because of its title, his book *The Education of a Gardener* is one I returned to again and again for several years running as we settled into our house in Long Island, where I dreamed of learning how to garden. Each year, I'd realize my timing was late, put the book back on its shelf only to pick it up again the following year.

ANNA PAVORD (1940–): Known throughout England for her columns on gardening in *The Independent*, Pavord is also a widely recognized name in America and elsewhere for her many books on plants and gardening. Her book *The Tulip: The Story of a Flower That Has Made Men Mad* was a bestseller. Pavord originally embarked on a writing career to finance the restoration of her home at the time, an eighteenth-century rectory and and garden in Dorset, England. The endeavor took thirty years.

ELEANOR PERÉNYI (1918–2009): Reading Perényi is like having a good friend with you in the garden, someone who is experienced and opinionated, who understands and forgives your mistakes, and who cheers your successes. An American by birth, she lived with her then husband in Hungary in the 1930s and chronicled her life in her memoir *More Was Lost*. She later moved to Connecticut, which became the setting for her book *Green Thoughts*. Comprised of seventy-two short essays arranged alphabetically—from Annuals and Artichokes, to Wildflowers and Woman's Place (in landscape gardening)—it's a highly original book. It was fun to discover that she appreciates earthworms perhaps even more than I do. This is one book I've gone back to again and again and read from cover to cover, learning more each time.

MICHAEL POLLAN (1955–): Pollan grew up in Long Island, then lived in Connecticut before moving to California. He wrote about building his Connecticut garden in *Second Nature: A Gardener's Education*. Since then, he's become a highly esteemed writer on everything from the intersection of nature and culture through agriculture and the food on your plate to farms and gardens. Several of his more recent books have been adapted for television. Pollan is interested in the essential qualities of a garden. *Second Nature: A Gardener's Education* is my favorite of his books. It examines humans' evolving relationship with nature and what it means to work with nature through one's garden.

VITA SACKVILLE-WEST (1892–1962): There are as many books written about Vita Sackville-West's colorful life and famous garden at Sissinghurst as there are books by her. Known for her outsized personality—she was the inspiration behind Virginia Woolf's character Orlando in the novel of the same name—Sackville-West was also a legendary gardener. Sissinghurst, the garden she created with her husband, Harold Nicolson, is one of the most famous gardens in England. I visited Sissinghurst during my research trip to England and went immediately to her highly original and legendary white garden, which was vibrant, even on a cold, gray September day. Sackville-West wrote a gardening column for *The Observer* for years in which she tackled every manner of horticultural question in her signature friendly, unintimidating voice.

SIR ROY STRONG (1935–): Strong is an art historian, writer, and museum curator—he was director of both the Victoria and Albert Museum and the National Portrait Gallery—and was knighted in 1982. Despite these accomplishments, he has said that his greatest achievement has been the creation of his garden. Strong and his wife, Julia Trevelyan Oman, created the Laskett Garden—set in Herefordshire on the Welsh borders—from a bare field. His book *Garden Party: Collected Writings 1979–1999* is one of the first and most memorable gardening books I ever read. A collection of essays on all things gardening, the book is just as eclectic as its author, whimsically yet also meaningfully tackling subjects like history and politics via the gardener's lens. His book remains one of my very favorites.

RECOMMENDED READING AND VIEWING

The books mentioned in the short biographies in Literary Mentors in the Garden (page 225) are those I've loved. The Selected Bibliography on page 241 contains a wide range of titles and types of gardening books, all of which I also recommend.

Here are several books that were especially helpful to me when I first started this project. Of course, there are others equally as good. But some do require a bit more gardening knowledge before you can plunge in.

- Jamaica Kincaid, *My Garden (Book):*

- Beverley Nichols, *Down the Garden Path*

- Eleanor Perényi, *Green Thoughts: A Writer in the Garden*

- Michael Pollan, *Second Nature: A Gardener's Education*

- Sir Roy Strong, *Garden Party*

And while it's technically a coffee-table book, Bridget Elworthy and Henrietta Courtauld's *The Land Gardeners: Cut Flowers* is so beautiful and the text so interesting that I'd include it, too.

There are many wonderful gardening series on streaming sites. I enjoy Monty Don's many programs, especially his shows on French and Italian gardens. Of particular note: *Monty Don's French Gardens*, where he visits Le Jardin Plume and interviews the Quibels (see page 42), and also the PBS documentary *Five Seasons: The Gardens of Piet Oudolf*, directed by Thomas Piper.

Additionally, three memorable movies—each one different from the next—where the garden plays a central role—are quite moving. Based on the classic novel by Giorgio Bassani, published in 1962, Vittorio De Sica's 1970 film *The Garden of the Finzi-Continis* is a tragic story of unrequited love that centers around a young man and the daughter of a wealthy and aristocratic Jewish family, the Finzi-Continis, during the 1930s when Italy was on the brink of World War II. As local Jews begin to gather inside the family's lavish estate to escape the Fascists, the garden of the Finzi-Continis provides sanctuary—literally and figuratively—from the brutal world outside its walls.

Alan Rickman's 2014 *A Little Chaos* is about two landscape artists—André Le Nôtre and Sabine de Barra—who build a garden in King Louis XIV's palace at Versailles. Though the film is largely fictitious, it does celebrate the genius of Le Nôtre, one of the most influential landscape architects of the French formal garden. Many say that Le Nôtre was to French gardening as Capability Brown had been to landscape design throughout Britain. The film is a delightful look at some of his early gardening ideas—namely his love of patterns and order—and how the fictitious

Sabine tries to introduce some creative chaos to his life and garden design.

Joel Hershman's 2000 *Greenfingers* is loosely based on a true story about British prisoners who compete in and win a national gardening competition. While the subject may seem like no laughing matter, the film is surprisingly funny, heartwarming, and inspirational, speaking, as it does, to the inherent value gardens bring to our lives, offering a sense of freedom, beauty, and escape where there is none.

I was also inspired by the short videos on the Garden Museum's web site, gardenmuseum.org.uk. The museum is enchanting—a place I love to visit when in London.

ACKNOWLEDGMENTS

Working with an exceptional team of people to publish a book once is a great privilege; twice, a blessing; and three times, like finding the pot of gold at the end of the rainbow.

Jonathan Burnham, president and publisher of the Harper Division of HarperCollins; Lynn Nesbit, my agent; and Mary Shanahan, the book's designer, have been at the core of these endeavors since my first book was published in 2013. Jonathan's wisdom is deep yet matched by his thoughtfulness, humanity, and generosity. I don't believe a writer could have a more dream publisher.

Lynn Nesbit innately understands which ideas will work and brings them to fruition with heart and soul. She cares deeply for books, even more for their writers—one of the many reasons she's beloved by her writers and very much by me.

Mary Shanahan has been our guiding hand, always gently but ever so creatively, suggesting this or that, and immediately, I'd see that she was right. She has great sensitivity and has made this book look the best it can be.

The editorial team now includes several others who've also been indispensable to this project: Elizabeth Viscott Sullivan is an editor of great imagination and brainpower. Her enthusiasm is a joy to behold. Along with Lynne Yeamans, she's led the book's form from start to finish. Liz introduced me to two people who've also been fundamental to the book: William Abranowicz and Signe Bergstrom.

The quality of Bill's work is obvious; it jumps off the page. Not only can he see the big picture and bring it to life, but he also gets in the weeds, as it were, on the level of the plants, to find their inherently special characteristics.

Signe Bergstrom generously and wisely helped me get the book's text off the ground and has continually given me great guidance throughout the seasons.

I lucked out the day I discovered the beautiful and powerful illustrations of All the Way to Paris, a design studio based in Copenhagen, led by Petra Gendt and Tanja Vibe. Over drinks with Jonathan, Liz, and me in New York City, I knew we'd found sympathetic souls, and just the right people to add the dimension of the illustrations, which capture so much emotion.

Two other people have been essential in this project: Billy Norwich and Katherine Schiavone. I owe them my deepest gratitude. One day out of the blue, a few years ago now, Billy wrote to me to suggest I do a book on a literary, garden-related subject. While that original idea didn't happen, the strong root of this book was in it, and had Billy not suggested it, I highly doubt I would have come to this project. Billy's generosity of friendship is huge, this being just one example. It's been my good fortune that Katherine, a talented garden designer, helped me make my garden. She's guided me with knowledge and grace, always letting the decisions be mine. It's been a partnership of the loveliest kind for me.

As I've learned, a book, perhaps even more so than a garden, can't be made alone. My wholehearted thanks to:

Anna Wintour, who gave me my start in the creative world many years ago when I edited the "Living" section of *Vogue*. Anna is not only a highly talented editor, she's also a champion of so many people in worlds that extend far beyond fashion, helping them in whatever ways she can.

Miranda Brooks, who showed me how much having a garden adds to the quality of life. I enjoyed watching her work so much that, although not realistic, I was often tempted to ask to be her intern.

Gaye Parise. I love working in the garden with Gaye. She instructs and shares her wealth of knowledge with kindness and makes me want to learn more.

Eric Piasecki and Ngoc Ngo, who graciously helped me get the project off the ground with exquisite photographs.

Hannah Wood, the wonderful editor of my previous book, has been a kindred spirit and the best possible reader. She's helped me puzzle through some sticky points with great generosity.

I'm also grateful to Anne Bass and Amanda Foreman, whose enthusiasm for their gardens enhanced my interest. Also, my thanks to Caroline Dean, Carol Mack, Nicola Shulman, and Christopher Woodward, Page Dickey and Susan Morgenthau, and Melissa Hill, Ann Marie Kuder, Tasha Lutek, Shirley Lord Rosenthal, and Patricia Vanterpool, who, in their distinct ways, helped the garden to grow.

I'll always remember my husband Don's enthusiasm for this project. He believed in anything I took on and was always my biggest supporter. It was exciting to watch his interest in the actual garden develop. I have a dream image of us sitting in two chairs at the garden's corner having a glass of wine. While that won't happen, this project wouldn't have happened without him.

Our children, William and Serena, have taken over the role of being my cheerleaders. Their enthusiasm is contagious. I am fortunate beyond measure to be their mother.

This book is for them: to Don, William, and Serena, with love.

SELECTED BIBLIOGRAPHY

Agnew, Harriet. "Take a Look around Parisian Super-Chef Alain Passard's Medieval Château." *Financial Times*. October 2019. https://www.ft.com /content/98c46142-da02-11e9-8f9b-77216ebe1f17.

Arkell, Reginald. *Old Herbaceous: A Novel of the Garden*. London: Michael Joseph, 1950.

Arnim von, Elizabeth. *Elizabeth and Her German Garden*. Rahway, NJ: The Mershon Company, 1898.

_____. *The Solitary Summer*. London: MacMillan & Co., 1899.

Austin, Alfred. *Love's Widowhood and Other Poems*. London: Forgotten Books, 2017.

Bannerman, Isabel. *Scent Magic: Notes from a Gardener*. London: Pimpernel Press, 2019.

Bemelmans, Ludwig. *Madeline*. New York: Viking Press, 1960.

Buchan, Ursula. *Garden People: The Photographs of Valerie Finnis*. London: Thames & Hudson, 2007.

Burnett, Frances Hodgson. *The Secret Garden*. London: William Heinemann, 1911.

Čapek, Karel. *The Gardener's Year*. London: William Heinemann, 1929.

Chatto, Beth. "Sir Cedric Morris, Artist-Gardener." *Hortus Revisited: A Twenty-first Birthday Anthology*. London: Frances Lincoln, 2008.

Churchill, Winston. *Thoughts and Adventures*. London: Thornton Butterworth Limited, 1932.

Colette. *Flowers and Fruit*. Edited by Robert Phelps. New York: Farrar, Straus and Giroux, 1986.

Cooper, David E. *A Philosophy of Gardens*. New York: Oxford University Press, 2006.

Cooper, Guy, and Gordon Taylor. *Mirrors of Paradise: The Gardens of Fernando Caruncho*. New York: The Monacelli Press, 2000.

Courtauld, Henrietta, and Bridget Elsworthy. *The Land Gardeners: Cut Flowers*. London: Thames & Hudson, 2019.

De Botton, Alain, and John Armstrong. *Art as Therapy*. New York: Phaidon Press, 2013.

Dickey, Page. *Duck Hill Journal*. Boston: Houghton Mifflin, 1991.

_____. *Embroidered Ground: Revisiting the Garden*. New York: Farrar, Straus and Giroux, 2011.

_____. *Uprooted: A Gardener Reflects on Beginning Again*. Portland, OR: Timber Press, 2020.

Don, Monty. *Down to Earth*. London: Dorling Kindersley, 2017.

Don, Monty, and Sarah Don. *The Jewel Garden*. London: Hodder & Stoughton, 2004.

Fenton, James. *A Garden from a Hundred Packets of Seed*. London: Notting Hill Edition, 2013.

Fish, Margery. *We Made a Garden*. London: W. H. & L. Collingridge Limited, 1956.

Fisher, M. F. K. *How to Cook a Wolf*. New York: Duell, Sloan and Pearce, 1942.

Garmey, Jane (ed.). *The Writer in the Garden*. Chapel Hill, NC: Algonquin Books, 1999.

Giubbilei, Luciano. *The Art of Making Gardens*. London: Merrell Publishers, 2016.

Hardyment, Christina. *Pleasures of the Garden: A Literary Anthology*. London: British Library, 2014.

Harrison, Robert Pogue. *Gardens: An Essay on the Human Condition*. Chicago: University of Chicago Press, 2008.

Hatch, Peter. *A Rich Spot of Earth: Thomas Jefferson's Revolutionary Garden at Monticello*. New Haven, CT: Yale University Press, 2012.

Hesse, Hermann. *Wandering: Notes and Sketches*. Translated by James Wright. New York: Farrar, Straus and Giroux, 1972.

Hesser, Amanda. *The Cook and the Gardener: A Year of Recipes and Writings from the French Countryside*. New York: W. W. Norton, 1999.

Holden, Linda Jane. *The Gardens of Bunny Mellon*. New York: Vendome Press, 2018.

Howard, Elizabeth Jane. *Green Shades: An Anthology of Plants, Gardens and Gardeners*. London: Aurum Press, 1991.

Jarman, Derek. *Derek Jarman's Garden*. London: Thames & Hudson, 1995.

Jekyll, Gertrude. *Children and Gardens*. London: Offices of Country Life, Ltd., and New York: C. Scribner's Sons, 1908.

———. *The Making of a Garden*. Edited by Cherry Lewis. New York: Antique Collectors Club, 1984.

Johnson, Hugh. *In the Garden*. London: Mitchell Beazley, 2009.

Keen, Mary. *Creating a Garden*. Hoboken, NJ: John Wiley & Sons, 1996.

Kincaid, Jamaica. *My Garden (Book):* New York: Farrar, Straus and Giroux, 1999.

Lane Fox, Robin. *Thoughtful Gardening*. New York: Basic Books, 2010.

Lappin, Shira. "Diana Athill: One Hundred Years in Gardens." *Hortus: A Gardening Journal*, 32, No.3. Herefordshire: The Bryansground Press, 2018.

Leighton, Clare. *Four Hedges: A Gardener's Chronicle.* New York: The Macmillan Company, 1935.

Lively, Penelope. *Life in the Garden.* London: Penguin Random House, 2017.

Mendelson, Charlotte. *Rhapsody in Green: A Novelist, an Obsession, a Laughably Small Excuse for a Vegetable Garden.* London: Kyle Books, 2016.

Merwin, W. S. *Garden Time.* Townsend, WA: Copper Canyon Press, 2016.

Miró, Joan. *I Work Like a Gardener.* Paris: Société Internationale d'Art XX Siècle, 1964.

Mitchell, Henry. *The Essential Earthman: Henry Mitchell on Gardening.* Bloomington, IN: Indiana University Press, 1981.

Nichols, Beverley. *Down the Garden Path.* New York: Atheneum Publishing, 1932.

_____. *Garden Open Today.* New York: Dutton, 1963.

_____. *Garden Open Tomorrow.* London: Heinemann, 1969.

_____. *Laughter on the Stairs.* London: Jonathan Cape, 1953.

_____. *Merry Hall.* London: Jonathan Cape, 1951.

_____. *A Thatched Roof.* New York: Doubleday, Doran and Co., Inc., 1933.

Page, Russell. *The Education of a Gardener.* New York: Random House, 1962.

Pavord, Anna. *The Curious Gardener.* London: Bloomsbury Publishing, 2010.

Pearson, Dan. *Natural Selection: A Year in the Garden.* London: Guardian Books and Faber & Faber, 2017.

Perényi, Eleanor. *Green Thoughts: A Writer in the Garden.* New York: Random House, 1981.

Pollan, Michael. *The Botany of Desire.* New York: Random House, 2001.

_____. *Second Nature: A Gardener's Education.* New York: Grove Press, 1991.

Potter, Beatrix. *The Peter Rabbit Library.* London: Frederick Warne & Co., 1902.

Quibel, Sylvie, Patrick Quibel. *Le Jardin Plume: Comme un jeu avec la nature.* Paris: Les Éditions Ulmer, 2017.

Richardson, Tim. *A Gardener's Year.* London: The Folio Society, 2008.

_____. *You Should Have Been Here Last Week: Sharp Cuttings from a Garden Writer.* London: Pimpernel Press Limited, 2016.

Rogers, Elizabeth Barlow. *Writing the Garden.* Boston: David R. Godine, 2011.

Sackville-West, Vita. *The Illustrated Garden Book: A New Anthology.* Edited by Robin Lane Fox. London: Michael Joseph, 1986.

_____. *In Your Garden.* London: Michael Joseph, 1951.

_____. *A Joy of Gardening.* New York: Harper, 1958.

Simonds, Merilyn. *Gardens: A Literary Companion.* Vancouver: Greystone Books, 2008.

Spender, Natasha. *An English Garden in Provence.* London: Harvill Secker, 1999.

Steinhauer, Jennifer. "Victory Gardens Were More About Solidarity Than Survival." *New York Times.* July 15, 2020.

Strong, Sir Roy. *Garden Party.* London: Frances Lincoln, 2000.

Stuart, Muriel. *Gardener's Nightcap.* London: Jonathan Cape, 1938.

Stuart, Rory. *What Are Gardens For?* London: Frances Lincoln, 2012.

Stuart-Smith, Sue. *The Well-Gardened Mind: The Restorative Power of Nature.* New York: Scribner, 2020.

Thaxter, Celia. *An Island Garden.* Boston: Houghton Mifflin & Co., 1894.

_____. *Among the Isles of Shoals.* Boston: James R. Osgood and Company, 1873.

Thoreau, Henry David. *Walden.* New York: Modern Library, 1992.

Verey, Rosemary. *A Countrywoman's Notes.* Boston: Little, Brown, 1989.

Warner, Charles Dudley. *My Summer in a Garden.* New York: Modern Library, 2002.

Welty, Eudora. "Pages Omitted from A Writer's Beginnings Notes and Tryouts," series 17, subseries 11, page 35. The State of Mississippi Department of Archives and History: the Eudora Welty Collection, Summer 1983. Copyright © Eudora Welty, LLC. Unpublished manuscript. PDF.

Wheeler, David (ed.). *Hortus Revisited: A Twenty-first Birthday Anthology.* London: Frances Lincoln, 2008.

White, Katharine S. *Onward and Upward in the Garden.* New York: Farrar, Straus and Giroux, 1979.

Wulf, Andrea. *Founding Gardeners: The Revolutionary Generation, Nature and the Shaping of the American Nation.* London: William Heinemann, 2011.

Young, Damon. *Voltaire's Vine and Other Philosophies.* London: Rider, 2014.

Frederick Childe Hassam, *The Room of Flowers*, 1894.

PHOTOGRAPHY AND ILLUSTRATION CREDITS

All the photography in this book is by William Abranowicz; all gouaches are by All the Way to Paris. All other art and photographs are credited as follows:

Unknown artist. 66, bottom left: *Dig for Victory, New Zealand,* c. 1943. Lithograph. © Imperial War Museums.

Unknown artist. 66, top right: *Paul J. Howard's California Flowerland,* c. 1940. Catalog cover. Los Angeles County Arboretum and Botanic Garden Collection.

Unknown artist. 134: Untitled [roses], date unknown. Pastel, 24¼ × 19¼ inches (48.25 × 37cm). Private collection.

af Klint, Hilma. 188: *Primordial Chaos No. 24,* 1906–1907. Oil on canvas, 20¼ × 14⁹/₁₆ inches (51.5 × 37cm). © The Hilma af Klint Foundation, Stockholm.

Bemelmans, Ludwig. 23: Illustration from *Madeline* by Ludwig Bemelmans, copyright 1939 by Ludwig Bemelmans, copyright renewed © 1967 by Madeleine Bemelmans and Barbara Bemelmans Marciano. Used by permission of Viking Children's Books, an imprint of Penguin Young Readers Group, a division of Penguin Random House LLC. All rights reserved.

Caillebotte, Gustave: 227: *The Gardeners,* 1875–77. Oil on canvas, 35½ × 46 inches (90 × 117cm). Private collection/Bridgeman Images.

Čapek, Josef. 79: Illustration from *The Gardener's Year* written by Karel Čapek. Prague: Aventinum, 1929.

Gibbs, Nathaniel K. 62: *In the Vegetable Garden, Monticello.* 2000. Oil on canvas. 12 x 16 inches (30.48 x 40.64 cm). Thomas Jefferson Foundation, Inc. © The Estate of Nathaniel K. Gibbs.

Grant, Duncan. 228: *The Doorway,* 1929. Oil on canvas, 34⅝ × 30½ inches (88 × 77.5cm). Arts Council Collection, Hayward Gallery, London. © Estate of Duncan Grant. All rights reserved, DACS, London/ARS, New York. Image: Bridgeman Images.

Hassam, Frederick Childe. 243: *The Room of Flowers,* 1894. Oil on Canvas, 34 × 34 inches (86.4 × 86.4cm). Private collection. Image © Universal Images Group North America LLC/Alamy Stock Photo.

Kelly, Ellsworth. 119: *Sunflower, 1983.* Graphite on paper, 22 × 30 inches (56 × 76cm). 137: *Rose, 1984.* Graphite on paper, 24⅛ × 18 inches (61 × 46cm). 154: *Tulip, 1984.* Graphite on paper, 24 × 17 inches (61 × 43cm). All © Ellsworth Kelly Foundation. Courtesy Matthew Marks Gallery.

Klimt, Gustav. 194: *Flower Garden,* 1907. Oil on canvas, 43¼ × 43¼ inches(110 × 110cm). Private collection. Image courtesy Belvedere Museum.

Le Bon, Mary. 66, bottom right: *Dig for Plenty,* 1944. Lithograph. © Imperial War Museums.

Leighton, Clare: 150: *Blackbird on Nest*. Wood engraving, $6^{15/16} \times 4^{7/8}$ inches (17.6 × 12.4cm). 151: *A lapful of windfalls*. Wood engraving, $6^{15/16} \times 4^{7/8}$ inches (17.6 × 12.4cm). Both engravings are from *Four Hedges: A Gardener's Chronicle*. New York: The Macmillan Company and London: Victor Gollancz, 1935. Courtesy the estate of Clare Leighton.

Llewellyn, Robert. 42–43: *View of Monticello Gardens*, 2010. Digital photograph. © Thomas Jefferson Foundation at Monticello.

Monet, Claude. 80–81: *The Artist's Garden at Giverny*, 1900. Oil on canvas, 32 × 36½ inches (81.6 × 92.6cm). Musée d'Orsay, Paris. © RMN-Grand Palais/Art Resource, NY.

Morley, Hubert. 66, top left: *Your Victory Garden Counts More than Ever!*, 1945. Lithograph. Image courtesy University of North Texas, UNT Digital Library.

Nicholson, Sir William. 170: *Miss Jekyll's Gardening Boots*, 1920. Oil on wood, 12¾ × 15¾ inches (32.5 × 40cm). Tate Britain. Digital Photograph. © Photo at Tate.

Piasecki, Eric. 97: © 2021 Eric Piasecki.

Potter, Beatrix. 22: Illustration from *The Tale of Peter Rabbit*. London: Frederick Warne & Co., 1902.

Schiavone, Katherine. 4: Watercolor of the garden plan, 2019. 101: Untitled, 2020. Drawing, graphite on paper, 13½ × 9¼ inches (34.29 × 23.5cm). 105: Untitled, 2020. Watercolor, 14 × 10¼ inches (35.5 × 26cm). Courtesy of the artist.

Tillmans, Wolfgang. 198: *untitled (jam)*, 2003. Image by Wolfgang Tillmans. Courtesy David Zwirner.

Twombly, Cy. 138–139: *Blooming*, 2001–2008. Acrylic, wax crayon on ten wooden panels, $98^{3/8} \times 196^{7/8}$ inches (249.9 × 500.1cm). © Cy Twombly Foundation. Photograph: Mike Bruce, courtesy Gagosian.

Becoming a Gardener
Copyright © 2022 by Catie Marron

HarperCollins books may be purchased for educational,
business, or sales promotional use. For information
please e-mail the Special Markets Department at
SPsales@harpercollins.com.

First published in 2022 by
Harper Design
An Imprint of HarperCollins *Publishers*
195 Broadway
New York, NY 10007
Tel: (212) 207-7000
Fax: (855) 746-6023

harperdesign@harpercollins.com
www.hc.com

Distributed throughout the world by
HarperCollins *Publishers*
195 Broadway
New York, NY 10007

ISBN: 978-0-06-296361-1
Library of Congress Control Number: 2021014977

Book design by Mary Shanahan
Front cover photograph by William Abranowicz

Printed in Italy
First Printing, 2022

ABOUT THE AUTHOR

Catie Marron's career has encompassed investment banking, magazine journalism, public service, and book publishing. She is the creator and editor of two anthologies published by HarperCollins, which explore the value and significance of urban public spaces: *City Squares: Eighteen Writers on the Spirit and Significance of Squares Around the World* (2016) and *City Parks: Public Places, Private Thoughts* (2013).

She is a trustee and chair emeritus of the New York Public Library, where she was chairman of the board from 2004 to 2011. Marron is also a trustee of Friends of the High Line, where she was also board chair, and a trustee of The Metropolitan Museum of Art and the Doris Duke Charitable Foundation.

Marron began her first career in investment banking at Morgan Stanley and then at Lehman Brothers. She then became senior features editor at *Vogue* and later a contributing editor for twenty years. While writing her books, Marron launched Good Companies, a curated, online guide to companies that strive to do good while also making a profit. This venture was shaped in part by the success of Treasure and Bond, a pop-up store that she cofounded with Nordstrom and Anna Wintour in 2011. All store profits went to charities benefiting New York City children.